CHANCING IT

By the same author

Unravelling the Mind of God:
Mysteries at the Frontiers of Science

25 Big Ideas:
The Science that's Changing our World

Why Don't Spiders Stick to their Webs?
And Other Everyday Mysteries of Science

CHANCING IT

The Laws
of Chance
and
How They
Can Work
For You

Robert
Matthews

P

PROFILE BOOKS

First published in Great Britain in 2016 by
PROFILE BOOKS LTD
3 Holford Yard
Bevin Way
London
WCIX 9HD
www.profilebooks.com

1 3 5 7 9 10 8 6 4 2

Typeset in Plantin by MacGuru Ltd
info@macguru.org.uk

Printed and bound in Great Britain by
Clays, Bungay, Suffolk

A CIP catalogue record for this book is available from the British Library.

ISBN 978 1 78125 030 3
eISBN 978 1 84765 862 3

FSC
www.fsc.org
MIX
Paper from
responsible sources
FSC® C018072

Contents

For Denise

The smartest person I know,
who unaccountably took a chance on me.

Introduction

One Sunday afternoon in April 2004, a 32-year-old Englishman walked into the Plaza Hotel & Casino in Las Vegas with his entire worldly possessions. They amounted to a change of underwear and a cheque. Ashley Revell had sold everything he owned to raise the $135,300 sum printed on the cheque; even the tuxedo he wore was hired. After exchanging the cheque for a depressingly small heap of chips, Revell headed for a roulette table, and did something extraordinary. He bet the lot on a single event: that when the little white ball came to rest, it would end up on red.

Revell's decision to choose that colour may have been impulsive, but the event itself wasn't. He'd planned it for months. He'd talked about it with friends, who thought it was a brilliant idea, and with his family, who didn't. Nor did some of the casinos; they may well have been fearful of going down in Vegas folklore as The Casino Where One Man Bet Everything And Lost. The manager of the Plaza certainly looked solemn as Revell placed the chips on the table, and asked him whether he was certain he wanted to go ahead. But nothing seemed likely to deter Revell. Surrounded by a large gathering of onlookers he waited anxiously as the croupier put the ball into the wheel. Then in one swift motion he stepped forward and put all his chips down on red. He watched as the ball slowed, spiralled in and bounced in and out of various pockets, and then came to rest – in pocket number 7. Red.

In that moment Revell doubled his net worth to $270,600. The crowd cheered, and his friends hugged him – and his father ruefully declared him 'a naughty boy'. Most people would probably take a harsher view of Revell's actions that day: at best ill-advised, certainly rash and possibly insane. For surely even billionaires for whom such sums are loose change would not have punted the lot on one bet. Would not any rational person have divided up such a sum into smaller wagers, to at least check whether Lady Luck was in town?

But here's the thing: having decided to do it, Revell had done precisely the right thing. The laws of probability show that there is no surer way of doubling your net worth at a casino than to do what he did, and bet everything on one spin of the wheel. Yes, the game is unfair: the odds in roulette are deliberately – and legally – tilted against you. Yes, there was a better than 50 per cent chance of losing the lot. Yet bizarre as it may seem, in such situations the best strategy is to bet boldly and big. Anything more timid cuts the chances of success. Revell had proved this himself in the run-up to the big bet. Over the previous few days he'd punted several thousand dollars on bets in the casino, and succeeded only in losing $1,000. His best hope of doubling his money lay in swapping 'common sense' for the dictates of the laws of probability.

So should we all follow Revell's example, sell everything we own and head for the nearest casino? Of course not; there are much better, if more boring, ways of trying to double your money. Yet one thing's for certain: they'll all involve probability in one of its many guises: as chance, uncertainty, risk or degree of belief.

We all know there are few certainties in life except death and taxes. But few of us are comfortable in the presence of chance. It threatens whatever sense we have of being in control of events, suggesting we could all become what Shakespeare called 'Fortune's fool'. It has prompted many to believe in fickle gods, and others to deny its primacy: Einstein famously refused to believe that God plays dice with the universe. Yet the very idea of making sense of chance seems oxymoronic: surely randomness is, by definition, beyond understanding? Such logic may underpin one of the great mysteries of intellectual history: why, despite its obvious usefulness, did a reliable theory of probability take so long to emerge? While games of chance were being played in Ancient Egypt over 5,500 years ago, it wasn't until the seventeenth century that a few daring thinkers seriously challenged the view summed up by Aristotle that 'There can be no demonstrative knowledge of chance'.

It hardly helps that chance so often defies our intuitions. Take coincidences: roughly speaking, what are the chances of a football match having two players with birthdays within a day of each other? As there are 365 days in a year and 22 players, one might put the chances at less than 1 in 10. In fact, the laws of probability reveal the true answer to be around 90 per cent. Don't believe it? Then check the birthdays of those playing in some football games, and see for yourself. Even then, it is hard to avoid thinking something odd is going on. After all, if you find yourself in a similar-sized crowd and ask whether anyone shares *your* birthday, you're very unlikely to find a match. Even simple problems about coin-tosses and dice seem to defy common sense. Given that a coin is fair, surely tossing heads several times on the trot makes tails more likely? If you're struggling to see why that's not true, don't worry: one of the great mathematicians of the Enlightenment never got it.

One aim of this book is to show how to understand such every-day manifestations of chance by revealing their underlying laws and how to apply them. We will see how to use these laws to *predict* coincidences, make better decisions in business and in life, and make sense of everything from medical diagnoses to investment advice.

But this is not just a book of top tips and handy hints. My prin-cipal goal is to show how the laws of probability are capable of so much more than just understanding chance events. They are also the weapon of choice for anyone faced with turning evidence into insight. From the identification of health risks and new drugs for dealing with them to advances in our knowledge of the cosmos, the laws of probability have proved crucial in separating random dross from evidential gold.

Now another revolution is under way, one which centres on the laws of probability themselves. It has become clear that in the quest for knowledge these laws are even more powerful than previously thought. But accessing this power demands a radical reinterpretation of probability – one which until recently provoked bitter argument. That decades-long controversy is now fading in the face of evidence that so-called Bayesian methods can transform science, technology and medicine. So far, little of all this has reached the public. In this book I tell the often astonishing story of how these techniques emerged, the controversy they provoked and how we can all use them to make sense of everything from weather forecasts to the credibility of new scientific claims.

Anyone wanting to wield the laws of probability must, however, know what their limitations are, and when they are being abused. It is now becoming clear that textbook methods long relied on by researchers to draw insights from data are routinely pushed far beyond their limits. Warnings about the potentially catastrophic consequences have circulated in academic circles for decades. Again, little of this emerging scandal has reached the public domain; this book seeks to remedy that. In doing so, it draws on my own contributions to the research literature, and includes ways of spotting when evidence and the methods applied to it are being pushed too far.

The need to understand chance, risk and uncertainty has never been more urgent. In the face of political upheaval, turmoil in financial markets and an endless litany of risks, threats and calamities, we all crave certainty. In truth, it never existed. But that is no reason for fatalism – or for refusing to accept reality.

The central message of this book is that while we can never be free of chance, risk and uncertainty, we now have the tools to take them on and win.

The coin-tossing prisoner of the Nazis

In the spring of 1940, John Kerrich set out from his home to visit his in-laws – no small undertaking, given that he lived in South Africa and they were in Denmark 12,000 kilometres away. And the moment he arrived in Copenhagen he must have wished he'd stayed at home. Just days earlier, Denmark had been invaded by Nazi Germany. Thousands of troops swarmed over the border in a devastating demonstration of blitzkrieg. Within hours the Nazis had overwhelmed the opposition and taken control. Over the weeks that followed, they set about arresting enemy aliens and herding them into internment camps. Kerrich was soon among them.

It could have been worse. He found himself in a camp in Jutland run by the Danish government, which was, he later reported, run in a 'truly admirable way'.[1] Even so, he knew he faced many months and possibly years devoid of intellectual stimulation – not a happy prospect for this lecturer in mathematics from the University of Witwatersrand. Casting around for something to occupy his time, he came up with an idea for a mathematical project that required minimal equipment but which might prove instructive to others. He decided to embark on a comprehensive study of the workings of chance via that most basic of its manifestations: the outcome of tossing a coin.

Kerrich was already familiar with the theory developed by mathematicians to understand the workings of chance. Now, he realised, he had a rare opportunity to put that theory to the test on a lot of simple, real-life data. Then once the war was over – presuming, of course, he outlived it – he'd be able to go back to university equipped not only with the theoretical underpinning for the laws of chance, but also hard evidence for its reliability. And that would be invaluable when explaining the notoriously counter-intuitive predictions of the laws of chance to his students.

Kerrich wanted his study to be as comprehensive and reliable as possible, and that meant tossing a coin and recording the result for as long as he could bear. Fortunately, he found someone willing to share the tedium, a fellow internee named Eric Christensen. And so together they set up a table, spread a cloth on it and, with a flick of a thumb, tossed a coin about 30 centimetres into the air.

For the record, it came down tails.

Many people probably think they could guess how things went from there. As the number of tosses increases, the well-known Law of Averages would ensure that the numbers of heads and tails would start to even out. And indeed, Kerrich found that by the 100th toss, the numbers of heads and tails were pretty similar: 44 heads versus 56 tails.

But then something odd started to happen. As the hours and coin-tosses rolled by, heads started to pull ahead of tails. By the 2,000th toss, heads had built up a lead of 26 over tails. By the 4,000th toss, the difference had more than doubled, to 58. The discrepancy seemed to be getting bigger.

By the time Kerrich called a halt – at 10,000 tosses – the coin had landed heads-up 5,067 times, exceeding the number of tails by the hefty margin of 134. Far from disappearing, the discrepancy between heads and tails had continued to grow. Was there something wrong with the experiment? Or had Kerrich discovered a flaw in the Law of Averages? Kerrich and Christensen had done their best to rule out biased tosses, and when they crunched the numbers, they found the Law of Averages had not been violated at all. The real problem was not with the coin, nor with the law, but with the commonly held view of what it says. Kerrich's simple experiment had in fact done just what he wanted. It had demonstrated one of the big misconceptions about the workings of chance.

Asked what the Law of Averages states, many people say something along the lines of 'In the long run, it all evens out'. As such, the law is a source of consolation when we have a run of bad luck, or our enemies seem on the ascendant. Sports fans often invoke it when on the receiving end of anything from a lost coin-toss to a bad refereeing decision. Win some, lose some – in the end, it all evens out.

Well, yes and no. Yes, there is indeed a Law of Averages at work in our universe. Its existence hasn't merely been demonstrated experimentally; it's been proved mathematically. It applies not only in our universe, but in every universe with the same rules of mathematics; not even the laws of physics can claim that. But no, the law doesn't imply 'it all evens out in the end'. As we'll see in later chapters, precisely what it does mean took some of the greatest mathematicians of the last millennium a huge amount of effort to pin down. They still argue about the law, even now. Admittedly, mathematicians often demand a level of precision the rest of us would regard as ludicrously pedantic. But in this case, they are right to be picky. For knowing precisely what the Law of Averages says turns out to be one of the keys to understanding how chance operates in our world – and how to turn that understanding to our advantage. And the key to that understanding lies in establishing just what we mean by 'It all evens out in the end'. In particular, what, exactly, is 'it'?

This sounds perilously like an exercise in philosophical navel-gazing, but Kerrich's experiment points us towards the right answer. Many people think the 'it' which evens out in the long run is the raw numbers of heads and tails.

So why did the coin produce far more of one outcome than another? The short answer is: because blind, random chance was acting on each coin-toss, making an exact match in the raw numbers of heads and tails ever more unlikely. So what happened to the Law of Averages? It's alive and well; the thing is, it just doesn't apply to the raw numbers of heads and tails. Pretty obviously, we cannot say how individual chance events will turn out with absolute certainty. But we can say something about them if we drop down to a slightly lower level of knowledge – and ask what chance events will do *on average*.

In the case of the coin-toss, we cannot say with certainty when we'll get 'heads' or 'tails', or how many we'll get of each. But given that there are just two outcomes and they're equally likely, we can say they should pop up with equal *frequency* – namely, 50 per cent of the time.

And this, in turn, shows exactly what 'it' is that 'evens out in the long run'. It's not the *raw numbers* of heads and tails, about which we can say nothing with certainty. It is their *relative frequencies*: the number of times each pops up, as a proportion of the total number of opportunities we give them to do so.

This is the real Law of Averages, and it's what Kerrich and Christensen saw at work in their experiment. As the tosses mounted up, the relative frequencies of heads and tails – that is, their numbers divided by the total number of tosses – got ever closer. By the time the experiment finished, these frequencies were within 1 per cent of being identical (50.67 per cent heads versus 49.33 per cent tails. In stark contrast, the raw numbers of heads and tails grew ever farther apart (see table).

No. of tosses	No. of heads	No. of tails	Difference (heads/tails)	Frequency of heads
10	4	6	–2	40.00%
100	44	56	–12	44.00%
500	255	245	+10	51.00%
1,000	502	498	+4	50.20%
5,000	2,533	2,467	+66	50.66%
10,000	5,067	4,933	+134	50.67%

The real Law of Averages, and what really 'all evens out in the end'

The Law of Averages tells us that if we want to understand the action of chance on events, we should focus not on each individual event, but on their relative frequencies. Their importance

Is a coin-toss really fair?

A coin-toss is generally regarded as random, but how the coin lands can be predicted – in theory, at least. In 2008, a team from the Technical University of Łódź, Poland,[2] analysed the mechanics of a realistic coin tumbling under the influence of air resistance. The theory is very complex, but revealed that the coin's behaviour is predictable until it strikes the floor. Then 'chaotic' behaviour sets in, with just small differences producing radically different outcomes. This in turn suggested that coin-tosses caught in mid-air may have a slight bias. This possibility has also been investigated by a team led by mathematician Persi Diaconis of Stanford University.[3] They found that coins that are caught do have a slight tendency to end up in the same state as they start. The bias is, however, incredibly slight. So the outcome of tossing a coin can indeed be regarded as random, whether caught in mid-air or allowed to bounce.

is reflected in the fact they're often regarded as a measure of that most basic feature of all chance events: their *probability*.

So, for example, if we roll a die a thousand times, random chance is very unlikely to lead to the numbers 1 to 6 appearing precisely the same number of times; that's a statement about individual outcomes, about which we can say nothing with certainty. But, thanks to the Law of Averages, we can expect the *relative frequencies* of the six different outcomes to appear in around 1/6th of all the rolls – and get ever closer to that exact proportion the more rolls we perform. That exact proportion is what we call the probability of each number appearing (though, as we'll see later, it's not the only way of thinking of probability). For some things – like a coin, a die or a pack of cards – we can get a handle on the probability from the fundamental properties that govern the various outcomes (the number of sides, court cards, etc.). Then we can say that, in the long run, the relative frequencies of the outcomes should get ever closer to that probability. And if they don't, we can start to wonder about why our beliefs have proved ill-founded.

↑UPSHOT

The Law of Averages tells us that when we know – or suspect – we're dealing with events that involve an element of chance, we should focus not on the events themselves, but on their relative frequency – that is, the number of times each event comes up as a proportion of the total number of opportunities to do so.

What the Law of Averages *really* means

The Law of Averages warns us that when dealing with chance events, it's their relative frequencies, not their raw numbers, we should focus on. But if you're struggling to give up the idea that it's the raw numbers that 'even out in the long run', don't beat yourself up; you're in good company. Jean-Baptiste le Rond d'Alembert, one of the great mathematicians of the Enlightenment, was sure that a run of heads while tossing a coin made tails ever more likely.

Even today, many otherwise savvy people throw good money after bad in casinos and bookmakers in the belief that a run of bad luck makes good luck more likely. If you're still struggling to abandon the belief, then turn the question around, and ask yourself this: why *should* the raw numbers of times that, say, the ball lands in red or black in roulette get ever closer as the number of spins of the wheel increases?

Think about what would be needed to bring that about. It would require the ball to keep tabs on how many times it's landed on red and black, detect any discrepancy, and then somehow compel itself to land on either red or black to drive the numbers closer together. That's asking a lot of a small white ball bouncing around at random.

In fairness, overcoming what mathematicians call 'The Gambler's Fallacy' means overcoming the wealth of everyday experiences which seem to support it. The fact is that most of our encounters with chance are more complex than mere coin-tosses, and can easily seem to violate the Law of Averages.

For example, imagine we're rummaging through the chaos of our sock drawer before racing off to work, looking for one of the few pairs of sensible black socks. Chances are the first few socks are hopelessly colourful. So we do the obvious thing and remove them from the drawer while we persist with our search. Now who says the Law of Averages applies, and that a run of coloured socks does not affect the chances of finding the black ones? Well, it may look vaguely similar, yet what we're doing is wholly different from a coin-toss or a throw of the roulette ball. With the socks, we're able to remove the outcomes we don't like, thus boosting the proportion of black socks left in the drawer. That's not possible with events like coin-tosses. The Law of Averages no longer applies, because it assumes each event leaves the next one unaffected.

Another hurdle we face in accepting the law is that we rarely give it enough opportunity to reveal itself. Suppose we decide to put the Law of Averages to the test, and carry out a proper scientific experiment involving tossing a coin ten times. That might seem a reasonable number of trials; after all, how many times does one usually try something out before being convinced it's true: three times, perhaps, maybe half a dozen? In fact, ten throws is nothing like enough to demonstrate the Law of Averages with any reliability. Indeed, with so small a sample we could easily end up convincing ourselves of the fallacy about raw numbers evening out. The mathematics of coin-tosses shows that with ten tosses it's odds-on that the number of heads and tails will be within 1 of each other; there's even a 1 in 4 chance of a dead heat.

Small wonder so many of us think that 'everday experience proves' it's the raw numbers of heads and tails that even out over time, rather than their relative frequencies.

↑UPSHOT

When trying to make sense of chance events, be wary of relying on 'common sense' and everyday experience. As we'll see repeatedly in this book, the laws ruling chance events lay a host of traps for those not savvy in their tricksy ways.

The dark secret of the Golden Theorem

Mathematicians sometimes claim they're just like everyone else; they're not. Forget the clichés about gaucheness and a penchant for weird attire; many mathematicians look perfectly normal. But they all share a characteristic that sets them apart from ordinary folk: an obsession with proof. This is not 'proof' in the sense of a court of law, or the outcome of an experiment. To mathematicians, these are risibly unconvincing. They mean absolute, guaranteed, *mathematical* proof.

On the face of it, a refusal to take anyone's word for anything seems laudable enough. But mathematicians insist on applying it to questions the rest of us would regard as blindingly, obviously true. They adore rigorous proofs of the likes of the Jordan Curve Theorem, which says that if you draw a squiggly loop on a piece of paper, it creates two regions: one inside the loop, the other outside. To be fair, sometimes their extreme scepticism turns out to be well founded. Who would have guessed, for example, the outcome of adding $1 + 2 + 3 + 4 +$ etc., all the way to infinity?[1] More often, a proof confirms what they suspected anyway. But occasionally a proof of something 'obvious' turns out both to be amazingly hard, and to have shocking implications. Given its reputation for delivering surprises, it's perhaps no surprise that just such a proof emerged during the first attempts to bring some rigour to the theory of chance events – and specifically, the definition of the 'probability' of an event.

What does '60 per cent chance of rain' mean?

You're thinking of taking a lunchtime walk, but you remember hearing the weather forecast warn of a 60 per cent chance of rain. So what do you do? That depends on what you think the 60 per cent chance means – and chances are it's not what you think. Weather forecasts are based on computer models of the atmosphere, and in the early 1960s scientists discovered such models are 'chaotic', implying that even small errors in the data fed in can produce radically different forecasts. Worse still, this sensitivity of the models changes unpredictably – making some forecasts inherently less reliable than others. So since the 1990s, meteorologists have increasingly used so-called ensemble methods, making dozens of forecasts, each based on slightly different data, and seeing how they diverge over time. The more chaotic the conditions, the bigger the divergence, and the less precise the final forecasts. Does that mean that a '60 per cent chance of rain at lunchtime' means 60 per cent of the ensemble showed rain then? Sadly not: as the ensemble is just a model of reality, its reliability is itself uncertain. So what forecasters often end up giving us is the so-called 'Probability of Precipitation' (PoP), which takes all this into account, plus the chances of our locality actually being rained on. They claim this hybrid probability helps people make better decisions. Perhaps it does, but in April 2009 the UK Meteorological Office certainly made a bad decision in declaring it was 'odds on for a barbecue summer'. To those versed in the argot of probability, this just meant the computer model had indicated that the chances were greater than 50 per cent. But to most everyone else, 'odds on' means 'very likely'. Sure enough, the summer was awful and the Met Office was ridiculed – which was always a racing certainty.

One of the most intriguing things about probability is its slippery, protean nature. Its very definition seems to change according to what we're asking of it. Sometimes it seems simple enough. If we want to know the chances of throwing a six, it seems fine to think of probabilities in terms of frequencies – that is, the number of times we'll get the outcome we want, divided by the total number of opportunities it has to occur. For a die, as each number takes up one of six faces, it seems reasonable to talk about the probability as being the long-term frequency of getting the number we want, which is 1 in 6. But what does it mean to talk about the chances of a horse winning a race? We can't run the race a million times and see how many times the horse wins. And what do weather forecasters mean when they say there's a 60 per cent chance of rain tomorrow? Surely it'll either rain or it won't? Or are the forecasters trying to convey their confidence in their forecast? (As it happens, it's neither – see box on previous page.)

Mathematicians aren't comfortable with such vagueness – as they showed when they started taking a serious interest in the workings of chance around 350 years ago. Pinning down the concept of probability was on their to-do list. Yet the first person to make serious progress with the problem found himself rewarded with the first glimpse of the dirty secret about probability that dogs its application to this day.

Born in Basle, Switzerland, in 1655, Jacob Bernoulli was the eldest of the most celebrated mathematical family in history. Over the course of three generations, the family produced eight brilliant mathematicians who together helped lay the foundations of applied mathematics and physics. Jacob began reading up on the newly emerging theory of chance in his twenties, and was entranced by its potential applications to everything from gambling to predicting life expectancy. But he recognised that there were some big gaps in the theory that needed plugging – not least surrounding the precise meaning of probability.[2]

Around a century earlier, an Italian mathematician named Girolamo Cardano had shown the convenience of describing chance events in terms of their relative frequency. Bernoulli decided to do what mathematicians do, and see whether he could make this definition rigorous. He quickly realised, however, that this seemingly arcane task created a huge practical challenge. Clearly, if we're trying to establish the probability of some event, the more data we have, the more reliable our estimate will be. But just how much data do we need before we can say that we 'know' what the probability is? Indeed, is that even a meaningful question to ask? Could it be that probability is something we can never know exactly?

Despite being one of the most able mathematicians of his age, it took Bernoulli 20 years to answer these questions. He confirmed Cardano's instinct that relative frequencies are what matter when making sense of chance events like coin-tosses. That is, he'd succeeded in pinning down the true identity of the 'it' in statements like 'It all evens out in the long run'. As such, Bernoulli had identified and proved the correct version of the Law of Averages, which focuses on relative frequencies rather than individual events.

But that wasn't all. Bernoulli also confirmed the 'obvious' fact that when it comes to pinning down probabilities, more data are better. Specifically, he showed that as data accumulate, the risk of the measured frequencies being wildly different from the true probability gets ever smaller (if you find this less than compelling, congratulations: you've spotted why mathematicians call Bernoulli's theorem the *Weak* Law of Large Numbers; the more impressive 'strong' version was only proved around a century ago).

In a sense, Bernoulli's theorem is a rare confirmation of a common-sense intuition concerning chance events. As he himself rather bluntly put it, 'even the most foolish person' knows that the more data, the better. But dig a little deeper, and the theorem reveals a typically subtle twist about chance: we can't ever 'know' the true probability with utter certainty. The best we can do is collect so much data that we cut the risk of being wildly wrong to some acceptable level.

Proving all this was a monumental achievement – as Bernoulli himself realised, calling his proof the *theorema aureum*: 'Golden Theorem'. He was laying the foundations of both probability and statistics, allowing raw data subject to random effects to be turned into reliable insights.

With his mathematician's predilection for proof satisfied, Bernoulli began collecting his thoughts for his magnum opus, the *Ars Conjectandi* – the Art of Conjecturing. Keen to show the practical power of his theorem, he set about applying his theorem to real-life problems. It was then that his *theorema* started to lose some of its lustre.

Bernoulli's theorem showed that probabilities can be pinned down to any level of reliability – given enough data. So the obvious question was: how much data was 'enough'? For example, if we want to know the probability that someone over a certain age will die within a year, how big a database do we need to get an answer that we can be sure is, say, 99 per cent reliable? To keep things clear, Bernoulli used his theorem to tackle a very simple question. Imagine a huge jar containing a random mix of black and white stones. Suppose we're told that the jar contains 2,000 black stones and 3,000 white ones. The probability that we'll pick out a white stone is thus 3,000 out of a total of 5,000, or 60 per cent. But what if we don't know the proportions – and thus the probability of picking out a white stone? How many stones would we need to extract in order to be confident of being pretty close to the true probability?

In typical mathematician's style, Bernoulli pointed out that before we can use the Golden Theorem, we need to pin down those two vague concepts 'pretty close to' and 'confident'. The first means demanding that the data get us within, say, plus or minus 5 per cent of the true probability, or plus or minus 1 per cent, or closer still. Confidence, on the other hand, centres on how often we achieve this level of precision. We might decide we want to be confident of hitting that standard nine times out of ten ('90 per cent confidence') or 99 times out of 100 ('99 per cent confidence'), or even more reliably.[3] Ideally, of course, we'd like to be 100 per cent confident, but as the Golden Theorem makes clear, in phenomena affected by chance such God-like certainty isn't achievable.

The Golden Theorem seemed to capture the relationship

between precision and confidence for the problem of randomly plucking coloured stones from not just one jar, but *any* jar. So Bernoulli asked it to reveal the number of stones that would have to be extracted from a jar in order to be 99.9 per cent confident of having pinned down the relative proportions of black and white stones it contains to within plus or minus 2 per cent. Plugging these figures into his theorem, he turned the mathematical handle ... and a shocking answer popped out. If the problem was to be solved by taken out stones at random, over 25,500 stones would have to be examined before the relative proportions of the two colours could be pinned down to Bernoulli's specifications.

This wasn't merely a depressingly large number, it was ridiculously large. It suggested that random sampling was a hopelessly inefficient way of gauging relative proportions, as even with a jar of just a few thousand stones, one would have to repeat the process of examining stones over 25,000 times to get the true proportion nailed down to Bernoulli's standard. Clearly, it would be far quicker simply to tip the stones out and count them. Historians still argue over what Bernoulli thought of his estimate;[4] disappointment seems to be the consensus. What is certain is that, after noting the answer, he added a few more lines to his great work – and then stopped. The *Ars Conjectandi* languished unpublished until 1713, eight years after his death. It's hard to avoid the suspicion that Bernoulli had lost confidence in the practical value of his Golden Theorem. It's known that he was keen to apply it to much more interesting problems, including settling legal disputes where evidence was needed to put a case 'beyond reasonable doubt'. Bernoulli seems to have expressed his disappointment in the implications of his theorem in a letter to the distinguished German mathematician Gottfried Leibniz, where he admitted he could not find 'suitable examples' of such applications of his theorem.

Whatever the truth, we now know that although Bernoulli's theorem gave him the conceptual insights he sought, it needed some mathematical turbocharging before it was fit for use in real-life problems. This was supplied after his death by the brilliant French mathematician (and friend of Isaac Newton) Abraham de

Moivre – allowing the theorem to work with far less data.[5] Yet the real source of the problem lay not so much in the theorem as in Bernoulli's expectations of it. The levels of confidence and precision he'd demanded from it may have seemed reasonable to him, but they turn out to be incredibly exacting. Even using the modern version of his theorem, pinning down the probability to the standards he set demands that around 7,000 stones be randomly chosen from a jar and their colour noted – which is still a huge amount.

It's odd that Bernoulli didn't do the obvious thing and rework his calculations with less demanding levels of precision and confidence. For even in its original form, the Golden Theorem shows this has a significant impact on the amount of data required; using the modern version, the impact is pretty dramatic. Taking Bernoulli's 99.9 per cent confidence level, but easing the precision level from plus or minus 2 per cent to 3 per cent, slashes the number of observations by more than half, to around 3,000. Alternatively, sticking with an error level of 2 per cent but reducing our confidence level to 95 per cent cuts the number of observations by even more, to around 2,500 – just 10 per cent of the amount estimated by Bernoulli. Do both – a bit less precision, a bit less confidence – and the figure plunges again, to around 1,000.

That's far less demanding than the figure reached by Bernoulli, though admittedly we've paid a price in terms of the reliability of our knowledge. Perhaps Bernoulli would have baulked at lowering his standards so far; sadly, we'll never know.

Today, 95 per cent has become the de facto standard for confidence levels in a host of data-driven disciplines, from economics to medicine. Polling organisations have combined it with the precision of plus or minus 3 per cent to arrive at their standard polling group size of 1,000 or so. Yet while they may be widely used, we should never forget that these standards are based on pragmatism, rather than some grand consensus of what constitutes 'scientific proof'.

↑UPSHOT

The dirty secret lurking in Bernoulli's Golden Theorem is that when trying to gauge the effects of chance, God-like certainty is unattainable. Instead, we usually face a compromise between gathering a lot more evidence, or lowering our standard of knowledge.

The First Law of Lawlessness

The true meaning of the Law of Averages has been mangled and misunderstood so badly and so often that experts in probability tend to avoid the term. They prefer arguably even less helpful terms like the Weak Law of Large Numbers – which sounds like an unreliable rule about crowds. So instead, let us break apart the Law of Averages into its constituent insights, and call them the 'Laws of Lawlessness'. The first centres on how best to think about events that involve an element of chance.

The First Law of Lawlessness

When trying to make sense of chance events, ignore the raw numbers. Focus instead on their relative frequency – that is, how often they occurred, divided by how often they had the opportunity to do so.

The First Law of Lawlessness warns us to be wary of claims based purely on raw numbers of events. That makes it especially useful when confronted by media coverage of, say, people with side effects to some new treatment, or lottery wins in a specific town. Such stories are typically accompanied by pictures of the tragic victims or lucky winners. There's no doubting the power of such stories. Even a single, shocking, real-life case can trigger historic changes in policy – as anyone who's been through airport security after 9/11 knows. And sometimes that's the appropriate response. But basing a decision on a handful of cases is usually a very bad idea.

The danger is that the cases appear to be typical, when in fact they're anything but. Indeed, the very fact they're so shocking is often because they're 'outliers' – the product of extremely rare confluences of chance.

The First Law of Lawlessness shows that we can avoid such traps by focusing instead on *relative frequencies*: the raw numbers of events, divided by the relevant number of opportunities for the event to occur.

Let's apply the law to a real-life example: the 2008 decision by the UK government to vaccinate pre-teen girls against HPV, the virus responsible for cervical cancer. This national programme was hailed as having the potential to save the lives of hundreds of women each year. Yet shortly after its launch, the media seemed to have compelling evidence that this was a dangerously optimistic view. They reported the tragic case of Natalie Morton, a fourteen-year-old girl who died within hours of being given the vaccine. The health authorities responded by checking stocks and withdrawing the suspect batch. This was not enough for some, however: they wanted the mass vaccination programme abandoned. Was this reasonable? Some would insist on invoking the so-called Precautionary Principle, which in its most unsophisticated form amounts to 'better safe than sorry'. The danger here lies in resolving one problem while creating another. Stopping the programme would eliminate any risk of death among its participants, but that still leaves the problem of how best to tackle cervical cancer.

Then there's the risk of falling for a trap that deserves to be much better known (and which we'll encounter again in this book). Logicians call it the '*Post hoc, ergo propter hoc*' fallacy – from the Latin for 'After this, therefore because of this'. In the case of Natalie's death, the trap lies in assuming that because she died *after* being vaccinated, the vaccination must have been the cause. Certainly, true causes always precede their effects, but reversing the logic has its dangers: people in car crashes typically put on seat belts before setting off, but that doesn't mean putting on seat belts causes crashes.

But let's assume the worst: that Natalie's death really was caused by a bad reaction to the vaccine. The First Law of Lawlessness tells us that the best way to make sense of such events is to focus not on individual cases, but instead on the relevant proportions. What are these? By the time of Natalie's death, 1.3 million girls had been given the same vaccine. That means the relative frequency of this kind of event was around 1 in a million. It was this that persuaded the UK government, in the face of protests from anti-vaccination campaigners, to resume the programme once the suspect batch had been withdrawn. This was the rational response if Natalie had indeed fallen victim to a rare reaction to the vaccine.

As it happens, this wasn't the case: the media had fallen into the trap of *post hoc, ergo propter hoc*. At the inquest into her death, it emerged that Natalie had a malignant tumour in her chest, and her death was unconnected to the vaccination. Even so, the First Law showed that the authorities had adopted the right approach by taking out just the suspect batch, rather than abandoning the whole programme.

Of course, the First Law isn't guaranteed to lead straight to the truth. Natalie could have been Case Zero of a reaction to the vaccine never seen during tests. And it was clearly right to look into the causes of the case for evidence that there could be more. The role of the First Law lies in preventing us being overly impressed by individual cases, and focusing our attention instead on relative frequencies, thus putting such cases in their correct context.

There are more general lessons here for managers, administrators and politicians determined to bring about 'improvements' following a handful of one-off events. If they ignore the First Law of Lawlessness, they risk taking action to deal with events that are exceedingly rare. Worse, having based the 'improvement' on a handful of cases, they may then decide to test it on a similarly small set of data, focus again on raw numbers rather than relative frequencies, and come to utterly erroneous conclusions. It could be anything from a spate of customer complaints to a staff suggestion about, say, a new way of doing things. They all tend to start with a few anecdotes which may or may not be significant. But the first step to finding out is to put them into their proper context – by turning them into the appropriate relative frequencies.

Sometimes making sense of events requires a *comparison* of relative frequencies. In the late 1980s, UK-based defence contractor GEC-Marconi became the focus of media coverage following a spate of over twenty suicides, deaths and disappearances among technical staff. Conspiracy theories started to emerge, fuelled by the fact that some of the victims were working on classified projects. While these made for intriguing stories, the First Law tells us to ignore the anecdotes and focus instead on relative frequencies – in this case, a comparison of the relative frequency of strange events at Marconi and those we'd expect within the general population. And that immediately focuses attention of the fact that GEC-Marconi was a huge company employing over 30,000 staff, and that the deaths had been spread over eight years. This suggests that the 'mysterious' deaths and disappearances may not have been so surprising, given the size of the company. That at least is what the subsequent police investigation concluded, though the conspiracy theories persist to this day.

In fairness, the importance of comparing relative frequencies is starting to catch on within the media. In 2010, France Telecom made headlines with a GEC-Marconi-like number of suicides: 30 between 2008 and 2009. The story flared again in 2014, when the company – now called Orange Telecom – saw a resurgence in suicides, with ten in just a few months. This time, the explanation *du jour* was work-related stress. But in contrast to the reporting of the GEC-Marconi cases, some journalists raised the key question prompted by the First Law: is the rate of suicides, rather than just the raw numbers, really all that abnormal – given that it's a huge company with around 100,000 employees?

That raises a tricky question that often emerges when trying to apply the First Law, however: what is the appropriate relative frequency to use in the comparison? In the case of Orange Telecom,

The strange case of the Bermuda Triangle

The First Law is especially useful when trying to make sense of spooky claims and conspiracy theories. Take the notorious case of the disappearance of ships and aircraft over a patch of the western Atlantic known as the Bermuda Triangle. From the 1950s onwards, there have been countless reports that bad things happen to those who enter the triangular-shaped area between Miami, Puerto Rico and the eponymous island. Many theories have been put forward to explain the events, from UFO attacks to rogue waves. But the First Law of Lawlessness tells us to focus not on the raw numbers of 'spooky' disappearances (which may or may not have happened), but instead compare their relative frequency to what we'd expect from any comparable part of the ocean. Do that, and something amazing emerges: it's entirely possible that all the unexplained disappearances really did take place. That's because tens of thousands of ships and aircraft pass through this vast, 1 million square kilometres of sea and airspace each year. Even if you include all those weird tales of the unexplained, it turns out the Bermuda Triangle is not even in the top ten of oceanic danger zones. Certainly the hard-nosed actuaries at world-renowned insurers Lloyd's of London aren't fazed by the raw numbers of supposedly 'spooky' events in the region. They don't charge higher premiums for daring to venture into it.

is it the national suicide rate (which is notoriously high in France, at 40 per cent above the EU average), or something more specific, like the rate among specific age ranges (suicide is the principal cause of death among 25–34-year-olds in France) or perhaps socio-economic grouping? The jury is still out on the Orange Telecom case; while it may be a statistical blip, others insist workplace stress is the real explanation. It's entirely possible that the truth will never be known.

Whatever the reality, the First Law tells us where to start in making sense of such questions. It also makes a prediction: that anything that encompasses enough people – from a government health campaign to employment with a multinational – has the ability to generate headline-grabbing stories, backed up with compelling real-life anecdotes, that mean less than they seem.

Try it yourself. Next time you hear of some national campaign that is generally a good thing but can have nasty side effects for some people – such as a mass medication campaign – make a note of it, wait for the horror stories, and then put the First Law to work.

↑UPSHOT

Chance events can shock us by their apparent improbability. The First Law of Lawlessness tells us to look beyond the raw numbers of such events, and focus instead on their relative frequencies – which gives us a handle on the probability of the event. And if low-probability events can happen, they will – given enough opportunity.

What are the chances of *that*?

Sue Hamilton was doing some paperwork in her office in Dover in July 1992 when she ran into a problem. She thought her colleague Jason might know how to solve it, but as he'd gone home, she decided to call him. She found his phone number on the office noticeboard. After apologising for disturbing him at home, she began explaining her problem, but barely had she begun than Jason interrupted to point out that he wasn't at home. He was in a public phone box whose phone had begun to ring as he walked past, and he'd just decided to pick it up. Amazingly, it turned out that that number on the noticeboard wasn't Jason's home number at all. It was his employee number – which just happened to be identical to the number of the phone box he was walking past at the moment she called.

Everyone loves stories about coincidences. They seem to hint at invisible connections between events and ourselves, governed by mysterious laws. And it's true. There are myriad invisible connections between us, but they're invisible primarily because we just don't go looking for them. The laws that govern them are also mysterious – but again, that's primarily because we rarely get told about them.

Coincidences are manifestations of the First Law of Lawlessness, but with a twist. That's because this law tells us what to do to make sense of chance events, while coincidences warn us of how difficult it can be to do this.

When confronted with an 'amazing' coincidence, the First Law tells us to start by asking ourselves about its relative frequency – that is, the number of times such an amazing coincidence could happen, divided by the number of opportunities such events have to occur. For a truly amazing coincidence, we'd expect the resulting estimate of the probability of the event to be astoundingly low. But as soon as we try to apply the law to coincidences such as Sue Hamilton's phone call, we run into trouble.

How do we even begin to estimate the number of such amazing events, or the number of opportunities they get to arise? What, for that matter, constitutes 'amazing'? This clearly isn't something we can define objectively, which in turn means we're on shaky ground insisting that we've experienced something inherently meaningful. The late, great, Nobel Prize-winning physicist Richard Feynman highlighted this common feature of coincidences with a typically down-to-earth example. During a lecture on how to make sense of evidence, he told his audience, 'You know, the most amazing thing happened to me tonight. I was coming here, on the way to the lecture, and I came in through the parking lot. And you won't believe what happened. I saw a car with the license plate ARW 357. Can you imagine? Of all the millions of license plates in the state, what was the chance that I would see that particular one tonight? Amazing!'

Then there's the awkward fact that we usually decide that a coincidence was 'amazing' only after we've experienced it, making our assessment of its significance entirely post hoc, and potentially very misleading. There's a Monty Python sketch based on the legend of William Tell that captures the dangers of post hoc rationalisation perfectly. It shows a crowd of people gathered round our eponymous hero, as he takes careful aim at the apple sitting on the head of his son – and hits it. The crowd duly cheers ... and we feel impressed too, until the camera pulls back to reveal Tell's son riddled with arrows from all the previous failed attempts to hit the apple. Tell's skill only appears amazing if we ignore all of these; likewise with coincidences. In reality, they are constantly occurring around us all the time, but the overwhelming majority are boring and insignificant. Every so often we'll spot something we decide is the equivalent of an arrow splitting an apple – and declare it surprising, amazing or even spooky, having studiously ignored the myriad less interesting events.

All this speaks to the fact that we humans are natural-born pattern seekers, prone to seeing significance in meaningless noise. Doubtless our cave-dwelling ancestors benefited from erring on the side of caution and hiding if something looked even vaguely like a predator. But this can all too easily slide into what psychologists called apophenia: a predilection for seeing patterns where none exists. We're all especially prone to one particular form of this, known as pareidolia. Every so often the media reports claims of 'miraculous' cloud formations, scorch marks on toast or features on Google maps that supposedly look like Christ, Mother Teresa or Kim Kardashian. And it's hard to disagree that they do. What we make of such 'miracles' depends on whether we think the chances of getting them by fluke alone are impossibly low. If we apply the First Law of Lawlessness, we have to confront the fact that the brain has myriad ways of making a face out of random swirl.

One of the most notorious cases of pareidolia centres on the so-called Face on Mars. In 1976, one of NASA's probes to the Red Planet sent back a picture that seemed to show the image of an alien on the Red Planet. It provoked controversy for 25 years, with most scientists dismissing it as nonsense. A few tried estimating the chances of getting so realistic a face by chance alone, but ended up mired in disputes over the figures they'd plugged into their relative frequency calculations. Finally, in 2001, the truth was revealed by sharp images taken by NASA's Mars Global Surveyor. These showed that the 'face' was indeed just a rocky outcrop, just as the sceptics had claimed.

When trying to make sense of a coincidence, it's easy to underestimate just how common such an 'amazing' event is – not least by defining how amazing it is only *after* seeing it, which is cheating, really.

How to predict coincidences

One of the most perplexing demonstrations of the laws of chance is the so-called Birthday Paradox: just 23 people are needed to give better that 50:50 odds that at least two will share a birthday. You don't need so big a group to demonstrate such coincidences, though: a random gathering of five people gives an evens chance that at least two will share the same star sign (or were born in the same month, if you're a rational Virgo and thus prefer a less silly example). The reason so few people are needed is because you're asking for *any* match between all the different ways of pairing the people in the group – which is surprisingly large: one can form 253 pairs from 23 people. This lack of specificity is key: if you demand an exact match with *your* birthday, you'll need a crowd of over 250 people to give better than 50:50 odds. Being less fussy and asking for a match of *any* two birthdays within a day either way hugely boosts the chances of a coincidence: indeed, there's a 90 per cent chance of finding such a 'near miss' coincidence among the players in any football match.[1]

⬆UPSHOT

Coincidences surprise us because we think they're very unlikely, and so can't be 'mere flukes'. The First Law of Lawlessness warns us of the dangers of underestimating the chances of coincidences by deciding ourselves what counts as 'amazing'.

Thinking independently is no yolk

In September 2013, John Winfield was in the kitchen of his home in Breadsall, Derbyshire, when he realised he needed some eggs. Popping out to the store, he returned with six, and began cracking them open. To his surprise, the first one had a double yolk – something he'd never seen before in his life. Then he cracked another, and saw another double yolker. Amazed, he carried on opening the eggs, and discovered every one had double yolks, including the final one – which he dropped on the floor in his excitement.

The amazing case of the six double yolkers was picked up by journalists, who helpfully did the maths to show how unlikely the event was. According to the British Egg Information Service, on average only around 1 in 1,000 eggs produced has a double yolk. And this prompted reporters to reach for their calculators plus some half-remembered notions about how to handle probabilities. They reckoned that if the chances of getting one double yolker was just 1 in 1,000, the chances of getting six must be 1 in 1,000 multiplied by itself six times, or 1 in a billion billion. That's an astronomical number: it implies that to witness what Mr Winfield saw just once, you'd have to have opened a box of eggs every second since the birth of the universe.

Some journalists twigged there was something dodgy about this reasoning, however. For a start, Mr Winfield was hardly the first since the Big Bang to report such an event. A quick trawl of the web revealed several similar reports, including an identical case of six double yolkers being found in Cumbria just three years earlier.

Science writer Michael Hanlon at the *Daily Mail* raised doubts about the 1-in-1,000 figure used in the calculation.[1] He pointed out that the chances of getting multiple yolkers depend heavily on the age of the hens, with younger ones being over ten times more likely to produce them. So while the 1-in-1,000 figure might be true on average, the double-yolker rate for farms with younger birds could easily be 1 in 100 – boosting the chances of getting a six-pack from such farms at least a million-fold.

Yet that can't be the whole explanation, as it still leaves the chances of getting six double yolkers at around 1 in 1,000 billion. Each year the equivalent of around 2 billion half-dozen packs are consumed in the UK, so even with the hugely increased chances, we'd still expect to hear of around two cases per millennium, not two in barely three years. When a calculation gives as crazily incorrect an answer as this, it's a sign there's something fundamentally wrong with the assumptions behind it. And the big assumption made in this one is that the probabilities of each event occurring separately really can be multiplied together. The laws of probability show that's only permissible if the events in question – in this case, the discovery of double yolkers – are independent of one another, so that we don't have to correct for any outside influence.

The notion that events are independent runs deep in the theory of chance events. Many 'textbook' manifestations of chance – repeated tosses of a coin, say, or throws of a die – are indeed independent; there's no reason to suspect one such event should influence any other. Yet while the assumption of independence keeps the maths simple, we must never lose sight of the fact that it's just that: an assumption. Sometimes it's an assumption we can safely make – as when trying to make sense of cricketer Nasser Hussain's legendary run of 'bad luck' in 2001 when he lost the toss fourteen times on the trot. While the chances of that are barely 1 in 16,000, there's no need to suspect anything strange; when one thinks of how many top cricket captains have tossed coins over the decades, it's an event that was clearly going to happen one day. But all too often, the assumption of independence isn't remotely justifiable. We live in a messy, interconnected world shot through with connections, links and relationships. Some are the result of the laws of physics, some of biology, some of human psychology. Whatever the cause of the connections, blithely assuming they don't exist can lead us into trouble. Indeed, so serious are the consequences that another Law of Lawlessness is merited:

The Second Law of Lawlessness

When trying to understand runs of seemingly 'random' events, don't automatically assume they're independent. Many events in the real world aren't – and assuming otherwise can lead to very misleading estimates of the chances of observing such 'runs'.

Applying the Second Law to the double-yolker story means thinking of ways in which finding one such egg in a pack might be linked to finding more. As we've seen, one way is that the contents of one box could have come from young hens, which are prone to producing double yolkers. Then there's the possibility of double yolkers being brought together by egg-box packers, increasing the chances of getting a box full of them. Again, that's known to occur: double yolkers tend to be relatively large eggs, and stand out among the otherwise small ones produced by young hens – and thus tend to get boxed up together. Some supermarkets even make a point of boxing up potential double yolkers together.

There are, therefore, solid grounds for thinking that finding one double yolker increases the chances of finding another in the same box – and thus for rejecting the idea of independence and the colossal odds that implies. Like the First Law, the Second Law has myriad uses – including making sense of seemingly spooky coincidences. Take the bizarre tale of how the *Titanic* disaster of April 1912 was foretold in eerily accurate detail by a book written fourteen years earlier. In the short story 'Futility', published in 1898, the American writer Morgan Robertson told the story of John Rowland, a deckhand aboard the largest ship ever built, which sinks with huge loss of life after striking an iceberg in the North Atlantic one April night. And the name of the ship? SS *Titan*. The parallels don't stop there, either. Robertson's vessel was, at over 240 metres in length, around the same size as the *Titanic*, was described as 'unsinkable', and carried fewer than half the lifeboats needed for those aboard. It was even struck on the same side: starboard.

This is certainly an impressive list of coincidences, and might lead one to wonder whether Robertson had based his book on a premonition. Maybe he did, but the smart money is on his plot-line being a demonstration of what coincidences emerge if events are not independent. When 'Futility' was published, a race to build colossal passenger ships was already well under way, driven by international competition to win the Blue Riband – the accolade awarded to the fastest Atlantic passenger liner. In the final decade of the nineteenth century, the largest vessels went from around 170 metres in length to well in excess of 200 metres – and 240 metres was patently not out of the question. As for what could wreak havoc on such leviathans, icebergs were already a recognised threat. So was the inadequate provision of lifeboats: there had been warnings that regulations had failed to keep pace with the rapid increase in the size of vessels. Clearly, correctly guessing the side hit by the iceberg was a simple 50:50 shot. Barely less surprising is Robertson's choice of name for his doomed ship. In the search for something evocative of a colossal vessel, SS *Titan* is clearly more likely to feature in a list of candidates than, say, SS *Midget*.

In short, Robertson's aim of penning a tragic but plausible tale about a doomed leviathan more or less compelled him to include events and characteristics not too far from those of the *Titanic*. A random choice simply wouldn't have made narrative sense.

↑UPSHOT

'Textbook' manifestations of chance, such as coin-tosses, can be assumed to be independent. But in the real world, that's often a dangerous assumption to make, even with runs of apparently rare events. The Second Law of Lawlessness warns against automatically assuming independence when estimating the chances of such a set of coincidences.

Random lessons from the lottery

S ince it began in 1988, Florida's state lottery has handed out over $37 billion in prizes, created over 1,300 dollar million-aires and put over 650,000 students through college. But on 21 March 2011, it turned a lot of Floridians into conspiracy theorists. After years of suspicion, that evening they believed they had finally been given proof of why they had never won anything despite years of trying: the whole lottery was a fix. Their evidence could hardly have been more impressive. Every evening, seven days a week, the lottery runs the Fantasy 5 draw, where 36 balls are put into a ran-domising machine and five winning balls are chosen at random. Or at least, that's what the organisers claim. But on that day in 2011, it was obvious the fix was in. As the balls popped up out of the machine, it was clear that the process was anything but random: the winning numbers were 14, 15, 16, 17, 18. Hard-core lottery players knew that the odds against winning the jackpot with any random pick were around 1 in 377,000, so clearly something very suspicious had happened.

In reality, something all too common had taken place: a dem-onstration that most of us have a less-than-perfect grasp of what randomness really is.

We all like to think we can learn from experience. And given how common random events are in our world, you'd think people would be pretty much up to speed with what randomness can toss their way. You could hardly be more wrong. Asked simply to define randomness, people typically mention characteristics like 'having no rhyme or reason' and 'patternless' – which isn't too bad, at least, up to a point. It's when they're asked to apply these intuitions to real-life problems that it all starts to go wrong.

In the 1970s, psychologist Norman Ginsburg at McMaster University, Canada, carried out studies to see how good people are at the seemingly simple task of writing down lists of 100 random digits. Most participants came up with well-jumbled sequences of digits, with few repeats digits, runs of consecutive numbers or any other numerical pattern. In other words, they did their best to ensure every digit got its 'fair share' of appearances in each otherwise patternless sequence. In the process, they inadvertently demonstrated a fundamental misconception about randomness.

It's true that there's no rhyme or reason to randomness: by definition it cannot be the outcome of any predictable process. It's also true that it is patternless. The problem is, that's something that's only guaranteed on huge (indeed, strictly speaking, infinite) scales. On every other scale, the lack of rhyme or reason of randomness is entirely capable of containing pattern-like sequences long enough to *seem* significant. Yet when asked to create some randomness of our own, we can't resist trying to reflect the patternless nature of infinite randomness in even the shortest burst of the stuff.

Clearly, what we need is regular exposure to short snatches of randomness, so we can get a feel for what it looks like on such scales. Fortunately, that's easily achieved – indeed, millions unwittingly do it worldwide several times a week. It's called watching lottery number draws on TV.

Many countries have national lotteries as a means of raising money for good causes. Most people tune in to watch the draws of lottery numbers simply to see whether they've won the jackpot – which, given that the odds are typically millions to one against, is usually an exercise in futility. Yet there's something to be said even for those who've not bought any lottery tickets tuning in occasionally, to see what randomness can do – and watch the numbers fall into what look suspiciously like patterns.

Many lotteries (including, until recently, the UK's national lottery) are '6-from-49'; that is, winning involves correctly guessing the six balls drawn from the 49 put into the randomising machine. This doesn't sound too difficult; it's oddly tempting to estimate that the chances of being able to guess the right set of six is 6 out of 49, or around 1 in 8. But like most forms of gambling (and that's what lotteries are), that's misleading, and the real chances are far lower. That 1 in 8 figure would be true if there were only six numbered balls among the 49, and we had to pick just one of the six. What we're being asked to do is far harder: pick all six of the right balls from 49, all of which have their own numbers on. The chances of doing this are very slim indeed: around 1 in 14 million. Why so small? Because our chances of getting the first number right are 1 in 49, the chances of getting the second number correct from the 48 remaining in the machine are 1 in 48; for the third they are 1 in 47 – and so on, all the way down to getting the sixth number right, which is 1 out of the 44 remaining. As the chance of any specific ball emerging from the machine is random and thus independent of the chances for any of the others, the probability of guessing all six of any given set is all these probabilities multiplied together – which is $(1/49) \times (1/48) \times (1/47) \times (1/46) \times (1/45) \times (1/44)$ – which works out at pretty much exactly 1 in 10 billion. The organisers of lotteries do cut us a bit of slack by not demanding that we also get the exact order in which the six come out of the machine correct too. They'll accept any of the 720 different orderings of six chosen balls (say, 2, 5, 11, 34, 41, 44, or 34, 2, 5, 11, 44, 41, etc.). So the odds of our picking the same numbers are around 1 in 10 billion times 720, which comes out at around 1 in 14 million. Just in case you think these aren't bad odds, picture this: they're equivalent to the lottery organisers tipping ten 1-kilogram bags of sugar on the floor, and asking you to pick out from the heap the single grain they've stained black – in one go, and while wearing a blindfold. Good luck with that.

So the chances are that we're never going to win the jackpot, even if we play for the rest of our lives. Indeed, one can show that the average lottery player in the UK has a bigger chance of dropping dead during the half-hour it takes to watch the show and make the call to claim the jackpot. But lurking in those numbers that routinely disappoint week in, week out, is an important lesson about the workings of randomness. Indeed, it's so important, it deserves elevation to the status of a Law of Lawlessness:

The Third Law of Lawlessness

True randomness has no rhyme or reason to it, and is ultimately patternless. But that doesn't mean it is devoid of all patterns at every scale. Indeed, on the scales on which we encounter it, randomness is shockingly prone to producing regularities which seduce our pattern-craving minds.

Evidence for this law can be found by regularly watching those lottery draws on TV – or, for those who need more rapid gratification, by looking through the online archives of previous jackpot selections. Examining a few weeks' worth of the winning six numbers for the UK national lottery at random (what else?) won't reveal any obvious patterns – seemingly confirming our belief that randomness really does mean patternless at every scale. For example, here's the eight sets of winning numbers in the UK draw in June 2014:

14, 19, 30, 31, 47, 48
5, 10, 16, 23, 31, 44
11, 13, 14, 28, 40, 42
9, 18, 22, 23, 29, 33
10, 11, 18, 23, 26, 37
3, 7, 13, 17, 27, 40
5, 15, 19, 25, 34, 36
8, 12, 28, 30, 43, 39

At first glance, it looks like 48 numbers with no obvious patterns, biases or sequences, just as we'd expect. But look again, this time for the most basic pattern possible with lottery balls: two consecutive numbers. Four of the eight sets contain such a 'run'; indeed, the first set contains two such runs. Chances are you missed them because they're such trivial patterns that they elude even the renowned pattern-spotting abilities of *H. sapiens*. Yet this is a hint of the patterns randomness can throw at us, and how they follow certain laws – all in seeming defiance of our beliefs about randomness. Using a notoriously tricky branch of maths called combinatorics, it's possible to count up the ways of getting runs of different lengths among the six numbers, and it turns out that *at least* two consecutive numbers can be expected in half of all '6-from-49' lottery draws. Thus in the eight draws during June 2014 we should expect around four to have a run of two or more numbers, and that's just what we got – and would get most months, if we'd bothered to check.

Before anyone thinks this might help predict which numbers are going to win each week, don't forget we still have no idea *which* two or more numbers will be paired: that's random, and thus unpredictable. What we've shown is simply that it will happen to *some* pair or longer run of numbers. Even so, this holds some very important lessons for us about patterns in randomness. First, it shows that patterns are not only possible in randomness, they're actually surprisingly common – and their rate of appearance can be calculated. Secondly, it highlights the fact that many samples of randomness – including lottery draws – have lots of patterns, but we miss them because we deem them 'insignificant'. In other words, we need to be wary about seeing 'significant' patterns in randomness, because patterns are in the eye of the beholder. And thirdly, while being very *specific* about what we want from randomness slashes the chances of getting it (e.g. that single set of six jackpot-winning balls), being very *vague* (e.g. 'any consecutive pairs') greatly increases the chances.

We can put all this to work by looking for other patterns in those samples of randomness we see in lottery draws. Viewers of the 1,310th draw for the UK national lottery on 12 July 2008 were astonished to witness no fewer than four consecutive numbers emerge among the six balls drawn from the 49 put in the randomising machine: 27, 28, 29, 30. A month later, the lottery machine was churning out other patterns, this time of three consecutive numbers among the six: 5, 9, 10, 11, 23, 26. While more striking than mere pairs, these patterns are still surprisingly common – not least because we're not fussed which three- or four-ball runs make up the pattern. Combinatoric calculations show that even the startling run of four consecutive numbers should pop up on average around once in every 350 draws – so arguably the biggest surprise was why it had taken 1,300-odd draws to see the first (sure enough, there have been several since).

In the light of such insights, the appearance of a complete run of five consecutive numbers in Florida's Fantasy 5 lottery draw of 21 March 2011 should no longer seem all that shocking. Again, we're not demanding a specific set of numbers, and that makes it easier to achieve. Indeed, it's easy to do the sums to see this. Following the reasoning for the UK lottery, extracting any five balls from the 36 in the Florida Fantasy 5 draw in the right sequence is possible in around 45 million ways. Again, the organisers cut us some slack, and any of the 120 different orderings of five balls is acceptable as a win, so there are around 375,000 ways of matching the jackpot-winning balls. But of these, only some will be all consecutive: the first such set is {1, 2, 3, 4, 5}, then {2, 3, 4, 5, 6} all the way up to {32, 33, 34, 35, 36}. There are just 32 such consecutive sets, so the probability of the five numbers being consecutive is 32/375,000 = 1 in 12,000. As draws are held seven days a week all year round, that means we should expect a roughly 30-year gap between each example of five consecutive balls. Give randomness enough time, and it'll come up with anything. In the event, the first popped up after 23 years, which is a little early, but not egregiously so.

There's one more valuable lesson about randomness we can learn from lottery draws – and a case study popped up in the UK lottery midweek draw shortly after that set of four consecutively numbered balls. First came a triple of 9, 10, 11, then another – 32, 33, 34 – the following week, and then another – 33, 34, 35 – the week after that.

This time we have a cluster of patterns. So what are we to make of this? Nothing – apart from its startling demonstration of how real randomness can throw up such clusters. Combinatoric calculations show that, in the long run, such triples will pop up in one in 26 draws from this kind of lottery. But randomness, with its customary lack of rhyme or reason, has no way of rigidly sticking to that rate. Sometimes the triples will be widely spaced, and sometimes they'll come in clusters as they did in 2008. Only conspiracy theorists are likely to see anything in such clusters. It's a different matter when the patterns produced by randomness represent not lottery numbers but, say, cancer cases in a town. Maybe there's something in the patterns, maybe there isn't, but even then we must remember that randomness is capable of producing patterns and clusters of patterns with surprisingly ease.

Sometimes the lottery does things that can make even mathematicians smile. After churning out the simple patterns in July and August, on 3 September 2008 the UK lottery spat out its most sophisticated pattern yet: 3, 5, 7, 9, four consecutive odd numbers. And after that, it went back to months of doing what randomness is 'supposed' to do: being dull, flat and patternless.

Many mathematicians regard playing lotteries as unspeakably stupid. They point to the shockingly low odds of winning the jackpot (remember those ten bags of sugar and that single grain?) and to the fact that the organisers set up lotteries so that players typically have to spend more than the average jackpot in tickets to stand a decent chance of winning. Which is true, though one could argue that paying for a ticket boosts the chances of winning by an infinite amount, from zero to 1 in 14 million, which is a lot of bang for your buck. But as we've seen, while you do have to be 'in it to win it', some priceless lessons about randomness can be had from any lottery for free.

⬆UPSHOT

Most of us think we know what randomness looks like: nice and smooth and utterly lacking in any patterns or clusters. The reality is very different – as the numbers that emerge during lottery draws show. They feature all kinds of patterns and clusters. But while the frequency of these patterns can be predicted, their precise identity never can.

Warning: there's a lot of X about

In May 2014, a verdict of suicide was recorded on a sixteen-year-old who asphyxiated himself in his bedroom in Hale, Greater Manchester. William Menzies was a straight-A student with no obvious problems. But the coroner had noticed something that worried him – something that connected the tragedy to another case of teenage suicide he'd personally dealt with, along with two others he'd encountered. All the victims had killed themselves after playing a video game. And not just any video game, either, but the best-selling *Call of Duty*, in which players take part in virtual warfare.

Among its millions of fans – and its critics – *Call of Duty* is known for its immersive realism. The notorious lone terrorist Anders Breivik claimed to have used it for training before slaughtering 77 people in Norway in one day in July 2011. Could it be that *Call of Duty* is so realistic it triggers the same side effects as real-life combat, such as post-traumatic stress disorder, depression and even suicidal thoughts? The coroner was sufficiently concerned of the risk that he issued a warning, urging parents to keep their kids away from such games.

Not everyone was convinced by his logic. Among the sceptics was Dr Andrew Przybylski, an experimental psychologist at the Oxford Internet Institute. He pointed out that millions of teenagers play *Call of Duty* in the UK, so it should hardly be a surprise if some of those who commit suicide also own it. Dr Przybylski underlined his point by an analogy: lots of teenagers wear blue jeans, making it pretty likely that many of those who commit suicide were wearing blue jeans at the time. So does it make sense to conclude blue jeans lead to suicide?

Put like that, it's clear why such arguments don't really stack up. First, they focus on only part of what's needed to make the case of a causal link between X and Y. That is, they focus on the surprisingly high probability of teenagers who commit suicide having recently played *Call of Duty*. But how do we *know* it's surprisingly high? The only way to tell is by putting it into context – which means comparing it to the probability of teenagers who *don't* commit suicide having recently played *Call of Duty*. And if we're dealing with something as ubiquitous as teenagers playing *Call of Duty*, you can bet a high proportion of perfectly happy teenagers will have played it too.

This highlights a general point: be wary about believing X explains Y if X is very common. But the flip-side is also true: if some effect is very common, be wary of blaming its appearance on some specific cause – as if it's very common, it's likely to have multiple causes. A classic example of this has recently made headlines in connection with a major public health debate in the UK. Statins are cholesterol-lowering drugs, and they've been shown to reduce the risk of death among people with a relatively high risk of heart disease. This has led some medical experts to propose that even people with little or no extra risk should also take statins as a preventive measure. It's a proposal that has sparked a huge row among experts and patients alike. Some see it as a step towards the 'medicalization' of the population, in which we all pop pills rather than live healthier lives. But most concern centres on the widespread reports of fatigue and muscular aches and pains among those taking statins. No one is dismissing the distress such symptoms bring – though some would argue they're a small price to pay in return for a reduced risk of premature death. What no one could argue with, however, is the fact that such symptoms are extremely widespread. And that leads to the suspicion that the link with statins may be entirely spurious.

This possibility has recently been put to the test by an analysis of studies collectively involving over 80,000 patients.[1] These studies were 'double blinded', meaning that neither the patients *nor* the researchers knew who was getting statins and who was getting a harmless placebo. The data showed that around 3 per cent of people given statins did indeed suffer from fatigue, and a startling 8 per cent from muscle aches. All very worrying – until one learns that virtually identical proportions of those patients getting the placebo also experienced these two symptoms. In other words, there's no reason to think that taking statins leads to an increased risk of their most 'notorious' side effects. They're simply so common that there's a relatively high chance that someone who starts taking statins will also experience an outbreak of fatigue and aches and pains – and, entirely understandably, blame the drugs.

Understandable, perhaps – but justifiable only if one has ruled out the risk of having mistaken ubiquity for causality. And sometimes it takes full-blown scientific studies involving huge amounts of data to do that.

Strangely enough, an entire class of scientific studies has been justified on the basis of this kind of flawed reasoning. It concerns perhaps the most controversial issue in experimental science: the use of animals. There's no question that experiments on animals have been important in many areas of medicine, from surgery to cancer research. Nor can anyone doubt that such use of animals has provoked strong reactions from both the pro- and anti-vivisection camps. The resulting debate has been vociferous, even violent, with each side exchanging claims and counter-claims. But for those who support the use of animals, one claim has acquired almost talismanic power: that 'virtually every' medical achievement of the last century has depended in some way on animal research.

Despite being quoted by leading researchers and even the Royal Society, Britain's premier scientific academy, the justification for this statement is far from clear. The wording comes from a claim made in an anonymous article in a newsletter circulated by the American Physiological Society about twenty years ago. It contains not a single reference to back up its impressive assertion. Even so, the implication is clear: it's vital to continue with experiments on animals if scientists are to continue finding life-saving drugs. Yet like the supposed link between suicide and video games, this overlooks a key issue: the sheer ubiquity of animal experiments. Since the thalidomide tragedy of the 1950s, there's been a legal requirement that every new drug undergoes testing on animals before it is allowed to be tested on human volunteers, let alone released onto the general market. As a result, every drug – regardless of whether it works in humans or not – will have been tested on animals. The fact that all the most successful drugs have been tested on animals is thus merely a truism, and tells us nothing about the causal link between the use of animals and the advancement of medicine. Claiming that it does makes as much sense as claiming that the equally ubiquitous practice of wearing lab coats is crucial to medical progress. As such, the statement endorsed by the Royal Society (among many others) is essentially vacuous. It's important to stress, however, that this doesn't imply animal experiments are pointless. What it does mean is that scientists need hard evidence if they are to prove the value of animal experiments. Surprisingly little work has been done in this area, and what has been done is largely not fit for purpose.[2] What evidence there is points to a rather more nuanced view of animal experiments than either side of the debate seems willing to concede. It suggests that animal models do have some value in detecting toxicity before human trials, but are poor indicators of safety. Put more prosaically, if Fido the dog reacts badly to some compound, it's likely humans will too. But if Fido can cope with it just fine, that says very little about what will happen in poor, delicate us.

↑UPSHOT

Proving that one thing causes something else is often tricky – and it's fraught with danger if either the suspected cause or effect is very common. Showing that the suspected cause always precedes the effect is a start, but in such cases it's rarely enough.

Why the amazing so often turns ho-hum

We see it everywhere, from breakthrough movies whose sequels suck to soaring stocks that suddenly collapse. Today's skyrocketing successes have a habit of turning into tomorrow's damp squibs. What's especially galling is the way they so often lose their magic at the very moment we notice them. Our friends tell us of some absolutely *amazing* local restaurant they visited last week, so we give it a go – and it's just ho-hum. We bet on a tennis player making headlines for her stellar performances – only to see her sink back into the pack. Sometimes it's hard not to think everything's just hype, and that most things are just, well, pretty average. And the thing is, when it comes to understanding this irksome quirk of life, you'd be on the right track.

Everyone's heard the line 'Don't believe the hype', which of course none of us would if only we could distinguish it from reliable judgements. Hype is usually taken to mean some kind of exaggeration of the truth, but that presumes we know what the truth actually is. This is where knowing a bit of probability theory helps. First, the Law of Averages tells us that when trying to gauge the typical performance of anything that can be affected by random effects, we should collect plenty of data. Clearly, it makes little sense to expect an amazing sequel from a first-time author or tyro movie director, as both have given us just one data-point on which to judge them.

But probability theory also warns us that collecting lots of data isn't enough; it must also be *representative*. By definition data solely about exceptional performance aren't representative. Yet that's exactly what we're being fed when we read rave reviews, see banner headlines, or hear pundits rave about some new soaring stock. As a result, when it comes to assessing exceptional events, we should always *fear the phenomenal*. Basing our judgement solely on evidence of exceptional performance makes us likely to fall prey to a tricksy effect known as regression to the mean. First identified almost 150 years ago by the English polymath Sir Francis Galton, it's still not as widely known as it should be, despite its ubiquity.

Perhaps the most common victims of regression to the mean are sports fans. They've seen it at work countless times, and may well have suspected something weird is going on – but rarely twig what it is. This is how it goes. At the start of the season, it all looks like business as usual – win some, lose some. Then the team goes off the boil, and starts heading for relegation. Action is clearly needed; heads must roll. After a run of defeats, the club gets the message and fires the manager. And sure enough, it does the trick: the team starts doing better under the new manager with his new tactics. But then it all starts going wrong again. After a run of solid performances, the team starts slipping. Just a few months after the upheaval, the team seems barely any better off – and the muttering about getting a new manager starts all over again.

This will sound familiar even to those who wouldn't know one end of a football from another. That's because the same phenomenon can be seen at work everywhere from underperforming schools to tanking stocks. The basic idea behind regression to the mean is not hard to understand. The performance of a team – or a school or stock price – depends on a host of factors: some obvious, some less so, but all of which contribute to the average or 'mean' level. Yet at any given time, the actual performance is unlikely to be dead-on average. It will usually be a bit above or below the mean level, as a result of nothing more significant than random variation. This can be surprisingly large, and persist for a surprisingly long time, but eventually its positive and negative impacts balance out, and the performance will 'regress' back to its mean value. The trouble is, regression to the mean is especially strong with the most extreme events, as these are typically the most unrepresentative of all. Anyone who acts on the basis of such extreme events alone risks falling victim to the cruellest part of regression to the mean: its ability to make a bad decision initially look like a good one.

So, for example, a manager brought in to run a sports team after 'compelling' evidence of poor performance may well benefit from a run of better performances. Yet the improvement may well be nothing more than regression to the mean, the team merely heading back towards its typical level of performance after the random bad run that cost the last manager his job. Wait long enough, and the typical level of performance will reassert itself. The first signs may well appear with players who seemed to sparkle under the new manager. They may just have had a run of luck that happened to coincide with the new manager's arrival, and so will also experience regression to the mean – and start to look more average as time wears on. Then the apparent boost enjoyed by the whole team starts to fade too. Of course, sometimes teams under-perform because managers genuinely lose their touch. Even so, research by statisticians and economists using real-life data shows that regression to the mean can and does affect sports teams, with managers hired and fired, but with little effect on overall team performance.

Once you know about regression to the mean, you'll start to see it everywhere. That's because we so often focus on extremes. Take management techniques aimed at boosting performance. Many line managers are convinced fear is the best motivator – and even claim to have hard evidence to prove it. Every time their team seriously underperforms, they call them in for a kicking – and sure enough performance improves. And don't give me all that stuff about rewarding performance, says the gung-ho manager: that's 'obvious' baloney. After all, when bonuses are given to the top sales team each quarter, it has a habit of becoming ho-hum next quarter; that's 'obviously' complacency.

And yes, the performance data do seem to prove it – unless you know about regression to the mean. The trouble is, gung-ho bosses rarely welcome being told that the 'compelling' evidence for their

The amazing curative powers of regression to the mean

In their quest for new therapies, medical researchers run the risk of being tricked into thinking they've found a miracle cure by regression to the mean. That's because, by its very nature, the search for such treatments often focuses on patients with abnormal characteristics, such as unusually high blood pressure. Yet sometimes these abnormalities can be nothing more significant than random deviations from normality which will fade away over time. Spotting such effects is a challenge for researchers testing a new drug, as they run the risk of being fooled into thinking the drug has brought about an improvement over time, when the condition has simply regressed to the mean. They deal with it by setting up so-called randomised controlled trials, in which patients are randomly allocated either to receive the drug, or receive a harmless placebo 'control'. As both groups are equally likely to experience regression to the mean, its effects can be cancelled out by comparing the relative cure-rates in both groups. Unfortunately, no such safeguards are available to us when a friend recommends some remedy for, say, back pain. Lacking any comparison group, it's hard to be sure that any benefit we get isn't just regression to the mean. Indeed, some doctors argue that patients who believe they've been cured by 'alternative medicine' such as homeopathy have benefited from nothing more than regression to the mean. Its advocates insist, however, that studies that take this possibility into effect have been carried out, and still show a net benefit.

effectiveness is probably nothing more than a statistical effect – which may be another reason so few know about it.

We can at least protect ourselves from self-delusion, however. For example, when it comes to making investments, we need to be very wary about the go-go stocks highlighted by financial pundits. They naturally focus on phenomenal, headline-grabbing performance – the classic breeding-ground for regression to the mean. Again, this isn't some theoretical risk. The Princeton University economist and scourge of Wall Street Dr Burton Malkiel has made a study of what happens to those who invest in 'obvious' winners.[1] He compiled a list of the equity funds that performed best over the five years 1990 to 1994. The top 20 of these funds outperformed the S&P 500 index by an impressive annual average of 9.5 per cent, and were 'obvious' winners. Malkiel then looked at how these same funds did over the next five years. Collectively, they underperformed by an average of more than 2 per cent relative to the whole stock market. The rankings of the top three slipped from 1st to 129th, 2nd to 134th and 3rd to a truly dismal 261th. Such is the power of regression to the mean to hand out lessons in humility.

As with football managers, however, a handful of investment managers really do seem to know what they're doing and achieve consistently impressive performance that can't be dismissed as some statistical fluke. One such is former Wall Street legend Peter Lynch, whose Magellan Fund performed astoundingly well during the 1970s and 1980s. Unfortunately, the evidence suggests that most 'star' fund managers are just temporarily benefiting from regression to the mean, and are destined to fade after a few years – and take our investments down with them.

↑UPSHOT

When it comes to making decisions based on performance, fear the phenomenal. Exceptional performance is, by definition, anything but representative. And that makes it especially likely to disappoint, courtesy of the Great Leveller that is regression to the mean.

If you don't know, go random

During a press conference in February 2002, US Defense Secretary Donald Rumsfeld was asked about the risk of the Iraqi dictator Saddam Hussein supplying terrorists with weapons of mass destruction. Clearly irked by the question, Rumsfeld famously replied as follows:

> *[A]s we know, there are known knowns; there are things we know we know. We also know there are known unknowns; that is to say we know there are some things we do not know. But there are also unknown unknowns – the ones we don't know we don't know.*[1]

It was a response that prompted shock and awe among Rumsfeld's critics. Some took it as proof positive that the Pentagon was under the control of a lunatic. Others regarded it as simply risible: the UK-based Plain Speaking Society awarded Rumsfeld a spoof prize for gobbledegook. A few, however, saw his response as a succinct statement of a disturbing truth about the reliability of knowledge: that there's ignorance, and then there's ignorance about one's ignorance. On the face of it, there's nothing we can do about the latter – for how can we protect ourselves against something we don't even know exists? In fact, there *is* something we can do to at least reduce the threat from unknown unknowns. Even more surprisingly, perhaps, it relies on randomness.

With its notorious lack of rhyme or reason, randomness would seem an odd source of security in the quest for knowledge. Yet this is precisely why it is so valuable: randomness embodies freedom from underlying assumptions, which is where our ignorance can manifest itself most destructively. This potent feature of randomness was brought to the attention of scientists principally through the efforts of one of the founders of modern statistics, whose name will pop up several times in this book: Ronald Aylmer Fisher. After graduating in mathematics from Cambridge University around a century ago, Fisher became fascinated with the challenge of drawing the most reliable insights from data – especially in the complex, messy life sciences. Working as a statistician at an agricultural research laboratory, Fisher devised a host of techniques for extracting insights from experiments beset by the unknown unknowns that plague such research – for example, variability in the fertility of soil. His textbook on the analysis of the results, *Statistical Methods for Research Workers*, published in 1925, became perhaps the most influential statistical book ever published. But chief among the tools he recommended was randomisation, which Fisher declared '… relieves the experimenter from the anxiety of considering and estimating the magnitude of the innumerable causes by which his data may be disturbed'.[2]

Nowhere has this advice been put to better use than in medicine, where it has proved vital in the quest for effective therapies. As early as the fourteenth century, the Italian scholar and poet Petrarch talked of testing new potions by getting 'hundreds or a thousand men' with identical characteristics, treating just half of them, and then seeing how they fared compared with those left untreated.[3] Since everything else about them is the same, any differences are likely due solely to the treatment. All very simple – except for one thing: what do we mean by 'identical' people? Ideally, they need to be identical to the typical patient likely to be given the treatment if it passes muster. The trouble is that people naturally have a host of differences: physical, emotional and genetic among others. The impact of these on the outcome creates a host of 'known unknowns'. Add in the unknown unknowns, and the method Petrarch described starts to look simplistic.

This is where randomness comes to the rescue. Instead of trying to cover everything that might possibly affect how people respond (and most likely failing), we get a sample of patients and randomly allocate them either to get the new therapy or be left untreated (or given a placebo treatment). Being a sample, it will never be perfect, but clearly it would be good to use as large a sample as possible. Petrarch himself mentioned that – but not the critical extra feature of randomness that Fisher recommended. By allocating patients completely at random, we reduce the risk of the sample being 'biased', accidentally or otherwise, towards those who might (or might not) benefit.

Having used randomness to solve the problem of 'identical' patients, we can then put the rest of Petrarch's account into action: creating two groups of patients, those in the so-called treatment arm, who get the therapy, and those in the control arm, who are given some comparison therapy (or perhaps just a placebo). It's entirely possible that one of the arms has more patients with, say, some unknown genetic trait that undermines the therapy. But by using many patients chosen at random, there's a good chance we've got pretty similar numbers of such patients in both arms. With the bias this might introduce thus mitigated, the assessment of the therapy becomes more reliable.

This isn't the only benefit from using randomisation, however. Once the results are in, they have to be interpreted correctly. For example, if a difference between patient groups does emerge, suggesting the therapy is effective, it's always possible it's just a fluke result. On the other hand, failing to find a difference could be the result of using too few patients. Quantifying the chances of such outcomes needs probability theory, and that's at its simplest and most trustworthy if one can assume there are no biases at work. Randomisation achieves this – and even helps deal with some tricky ethical questions. Unscrupulous researchers might want to give their drug to less sick patients, while the others get the less effective old therapy – thus boosting the chances of the new drug doing well. On the other hand, compassionate researchers may want to give the new therapy to patients who would otherwise have little hope ... but that would mean condemning other patients to receiving the less effective treatment. Or at least it would do, if researchers were good judges of the likely effectiveness of their therapies. They're not: a 2008 analysis of over 600 randomised controlled trials of cancer treatments deemed worth trying on patients by the US National Cancer Institute since the mid-1950s found that only 25 to 50 per cent proved successful.[4]

Such ethical dilemmas are avoided simply by insisting on random allocation to each arm – and, moreover, by someone else unconnected to the trial.

In 1947, the UK Medical Research Council decided to give the power of randomness a test in a pioneering study of the effectiveness of the antibiotic streptomycin against tuberculosis. It wasn't very large: around a hundred patients were randomly allocated to receive either the standard treatment of just staying in bed, or bed-rest plus the antibiotic. To avoid the doctors or patients biasing the outcome through knowing who was getting what, all were kept in the dark ('blinded') about the outcome of the random selection process. After six months, the results were in – and on the face of it, they were impressive: of the fifty or so patients given the antibiotic, the survival rate was almost four times that of their counterparts getting only bed-rest. The trial was small, yet statistical tests suggested so big a difference was unlikely to be a fluke. Today, such so-called 'blinded' Randomised Controlled Trials (RCTs) have become the gold standard in testing the effectiveness of new therapies. Hundreds of thousands have been carried out, some involving tens of thousands of patients, and the results have benefited the health of countless millions. All this bears testament to the ability of randomness to reduce the impact of ignorance – both known and unknown. Its success in medicine has prompted attempts to use the RCT method in other areas of research aimed at tackling such ills as poverty and youth crime (see box overleaf).

Yet for all its power, the RCT isn't the infallible guide to 'what works' that some seem to think it is. While randomness can in principle deal with any unknown unknown, in reality it runs into the problem that so much research is done *by* humans *on* humans. For example, it's easy to randomise people once they've been recruited

Giving government policy the randomness treatment

The success of Randomised Controlled Trials (RCTs) of drugs in determining 'what works' has led to interest in using the same idea in other areas – such as testing out government policy. Politicians have a reputation for launching grand schemes based on little more than anecdotes and hunches. Wouldn't it be better to put their ideas to the test, using randomisation to combat their assumptions of omniscience?

It's an attractive idea – at least, to those wedded to the idea that policy should be based on facts rather than dogma. Perhaps its biggest success to date has been Mexico's *Oportunidades* ('Opportunities') social welfare programme, which tackles poverty by giving cash to specific families in exchange for regular school attendance, medical checks and dietary support.[9] The idea of offering money as a quid pro quo for participation was dismissed as naive by critics, and so the government responded by having the idea tested using an RCT. Hundreds of villages were randomised to either take part or act as controls, and the impact on welfare monitored. Two years later, the impact was assessed – and the policy was found to be effective in boosting both the welfare and the future prospects of those taking part. In 2002 the programme was rolled out to urban communities as well, and has proved such a success it's being copied elsewhere – including New York City.

Not every politically inspired idea has benefited from the RCT treatment. Take the 'Scared Straight' policy of dealing with juvenile delinquents by letting them witness the horrors that await them if they end up in jail. Named after an eponymous 1978 US documentary, it suggested hoodlums mended their ways after being exposed to 'lifers' in a New Jersey prison. Some politicians called for its wider use, but fortunately not everyone was willing to mistake anecdote for evidence. The scheme was subjected to a series of RCTs, and when analysed in 2013, the results showed the policy was actually worse than useless: those taking part had *higher* rates of delinquency than those left alone.[10]

Happily, there are signs that some governments are starting to see that RCTs are a surer way of finding out 'what works' than gut feeling.[11]

by the experimenters – but what if the researchers only recruit certain types of people? Over the years, randomised studies have led psychologists to a host of insights into human nature. Yet the exigencies of cost, time and convenience mean that many of these insights come from randomised studies of distinctly non-random types of human: American psychology students. In 2010, researchers from the University of British Columbia, Canada, published an analysis of hundreds of studies published in leading psychology journals and found that over two-thirds of the participants came from the USA, and of those, two-thirds were psychology undergraduates. Worse still, the researchers found that these students are especially unrepresentative of 'typical' humans – being overwhelmingly from societies which are Western, educated, industrialised, rich and democratic (or, as they put it, WEIRD).[5] Biases can also rear their heads while an RCT is under way – for example, when only certain types of people prove able (or willing) to stick to some strict dietary programme. Who knows why they drop out; maybe it's random, maybe it isn't, but it could just undermine the 'external validity' of the results – that is, the extent to which they'll apply to you or me. The truth is, there are a host of ways in which everything from drugs to diet supplements can work fine in scientific studies, but fall flat in the real world.[6]

And those are just the RCTs we get to hear about. Not even randomness can protect us from so-called publication bias, in which research findings deemed inconclusive, boring or 'unhelpful' just never get published. Various studies have shown that positive results are more likely to get published than negative or unhelpful ones.[7] The causes are hotly debated. Some blame slack practices by researchers; others claim that journals are too keen on headline-grabbing findings. Pharmaceutical companies have been accused of burying negative results to protect their share price. What isn't in doubt is the potentially dire effect publication bias can have on attempts to answer key questions by pooling together the published evidence. The resulting 'meta-analysis' is likely to be overly optimistic, with potentially life-threatening consequences for the public.

Finally there's the problem of dodgy researchers. Randomness is powerless to counteract the bias introduced by researchers who set up an RCT specifically to reach the 'right' answer. RCTs overseen by pharmaceutical companies have been criticised for using 'straw man' designs, in which the new drug is compared to some inappropriately feeble remedy – thus boosting the chances of headline-grabbing results.[8]

Like all human creations, RCTs can be subverted in a host of ways. But their use of randomness ensures that, for all their flaws, they're still the best means we have of protecting ourselves from the delusion of omniscience.

↑UPSHOT

The very lawlessness of randomness makes it invaluable in cutting through misplaced assumptions – both conscious and unconscious – and questionable practices. But if used badly, or partially, it can make shoddy research look 'scientific'.

Doing the right thing isn't always ethical

Thinking of swapping artificial sweetners for sugar? Think again: it may increase your risk of diabetes. Worried about losing your job? You might soon have asthma to add to your woes. Taking sleeping pills because you're fretting about all these health threats? You may substantially increase your risk of Alzheimer's disease.

The list of supposed threats to our health seems to get ever longer; these latest additions reared their scary heads in the media over the course of just one month in 2014.[1] Yet it's often hard to know what to make of such stories. Many seem to be based on research carried out by reputable scientists and published in respected journals. But it doesn't help that the evidence on any specific health threat so often seems to swing this way and that. Some years ago, coffee was condemned for increasing the risk of pancreatic cancer. That risk faded away, and now it seems it may be good for liver cancer.[2]

Deciding what to do on the basis of one-off media reports alone clearly makes no sense. What's needed is a proper scientific evaluation – and how better to conduct it than via that gold standard for medical investigation, the Randomised Controlled Trial (RCT)?

Not so fast: such a trial would demand a random sample of volunteers, and then deliberately exposing half of them to some unknown, and potentially harmful, risk factor. That raises some obvious ethico-legal issues. But they're not the only problem with

RCTs. While it would doubtless be fascinating to know if, say, people who become vegetarians in later life are healthier than those who eat meat, it's going to be tough to recruit thousands of people and then tell half of them they can never eat meat again for the rest of their lives.

For all its advantages, the RCT simply cannot be used to investigate some questions – despite the fact that they're often among the most interesting one can ask. So instead researchers use the so-called observational study. As its name suggests, this typically involves observing two groups of people, comparing them in the search for evidence of the effect under study. Which doesn't sound so different from an RCT, except for the absence of its most potent feature: randomisation. Unable to call on its power to deal with unknowns (both known and otherwise), observational studies try a different approach. As we'll see, it's not easy to apply; indeed, the evidence suggests it's rarely done effectively. And that's one big reason so many media health scare stories seem to flip-flop this way and that. Most are based on results from observational studies – which all too often reveal their shortcomings as substitutes for RCTs.

The most common type of observational study has a so-called 'case-control' format, which is a relatively cheap but fast means of investigating a possible link between some medical condition and a putative risk factor. Case-control studies have spawned a host of headline-grabbing health stories, such as those alleged links between taking sleeping pills and developing Alzheimer's disease. Setting up such a study involves finding lots of people with the condition (the 'cases'), and then a matched group of ordinary people (the 'controls'). The two groups are then compared. What the researchers are looking for are signs that people afflicted by the condition also tend to be those with the higher exposure to the supposed cause.

The most obvious problem with this is getting a 'matched group'. Without randomisation, researchers are forced to decide what criteria to match the two groups according to. Include too many criteria, and you soon run out of enough controls to pair off against the cases; include too few, and the comparison becomes a joke. Pick the wrong matching criteria, and it's possible the real link will just vanish in the matching process. Add in the risk of bias in choosing who gets picked to join either group in the first place, and the scope for unreliable results becomes obvious.

For all their potential failings, however, case-control studies are often the only ethical way to investigate concerns about putative health risks – especially for uncommon diseases, where one would otherwise have to observe huge numbers of people to reach reliable conclusions. And they do have some spectacular successes to their name. The most famous is the evidence for a link between lung cancer and cigarette smoking, uncovered by a case-control study published in 1950 by two of the most celebrated names in medical statistics: Austin Bradford Hill and Richard Doll. Armed with over 1,000 cases and controls, they were able to take into account a host of potentially relevant factors, from age and sex to social class, forms of domestic heating and even exposure to other pollutants. The relative proportions of smokers and non-smokers among the cases of cancer and those free of the disease pointed to a hefty increase in lung cancer risk from smoking. They went farther, however, and showed that the risk increased with increased consumption – a 'dose-risk' relationship that's certainly consistent with smoking being a cause of lung cancer. It does not prove it, however: without randomness to combat at least some of the biases, there's a substantial risk that some 'unknown unknown' was really responsible. And then there was the problem that both the cases and controls had been hospital patients – who might be unrepresentative of the general population.

Doll and Hill responded by setting up another widely used means of investigating health effects: the prospective cohort study. This time, instead of looking back at what might have triggered the effect, a prospective study follows a large population – 'cohort' – of people without knowing whom will be affected. This time, the effect of 'unknown unknowns' is addressed by choosing a cohort of people alike in many ways – for example, having the same sex and socio-economic background. They will differ, however, according to whether they're exposed to the suspected cause of the effects being investigated.

Doll and Hill decided to focus on doctors, and by the early 1950s had managed to recruit a cohort of over 34,000 men and over 6,000 women, divided up into smokers and non-smokers. They then set about following the fate of the two groups in a study that lasted until 2001. What became known as the British Doctors Study found compelling evidence that cigarette smoking increased the risk of lung cancer roughly tenfold, and at least twentyfold in heavy smokers.

This unequivocal success encouraged researchers to turn to case-control and prospective studies to address a host of other health-related questions. This has resulted in such studies becoming a happy hunting ground for the media, which can rebut the charge of scare-mongering by pointing to the fact that the studies have been published by some 'prestigious' research journal or other. Among researchers themselves, however, the limitations of observational studies are causing increasing concern. Much of it surrounds the apparent failure of so many observational studies to reach any kind of consensus. The results of case-control studies in particular have become notorious for flip-flopping, with successive studies often failing to replicate earlier findings, or flatly contradicting them. One review of the use of such studies in linking ailments to specific genes found that out of 166 such links investigated multiple times, barely 4 per cent were consistently replicated.[3] Prospective cohort studies have generally fared better, but even the seemingly most impressive frequently fail to produce compelling conclusions.

Take the ongoing furore over the health implications of eating meat. In 2009, a huge cohort study of half a million Americans monitored over ten years revealed a clear link between eating red meat and the risk of cancer, cardiovascular disease and reduced longevity. Then in 2012, a huge Japanese study revealed no such risk, while in 2013 a huge European study came up with a mixed bag of results.[4] If even such huge observational studies, carried out by renowned experts, can't give consistent insights, what's the point of doing them? In fairness, it's possible that *both* studies are right. Differences in the composition of beef and how it's cooked and consumed (and indeed among those doing the cooking and consuming) could lead to American meat proving less healthy, at least for Americans. This again highlights the problem of generalisability that even RCTs can run into: the way the study was conducted can produce outcomes that apply only in special circumstances that aren't generally applicable.

Even so, these studies may well have fallen victim to the lack of randomisation that gives RCTs their power. To deal with this, the researchers tried to identify and cancel out ('control for') the impact of as many potentially misleading factors as they could, such as smoking history and alcohol intake. Doing this requires slicing and dicing the cohort data into lots of subgroups. And that means many of the findings are based on only a tiny fraction of the impressive-sounding half-million people making up the total cohort. Even then, it's possible that the outcomes of both studies were still subject to subtle biases. In 2011, two researchers from the US National Institute of Statistical Sciences cast light on the dangers of trying to mimic the benefits of RCTs by examining the claims made in observational studies that were subsequently tested against the 'gold standard' of an RCT. Of the 52 claims made in the twelve observational studies identified, the number confirmed by the subsequent RCT was … zero.[5]

So when faced with a report of some health risk or other (or benefit) uncovered by an observational study, how should we react? Epidemiologists, as those working in this area of research are known, often apply a few rules of thumb for deciding what findings are worth taking seriously (see box overleaf). This, in turn, has led to the emergence of a 'pecking order' when it comes to observational studies. The lowest form of epidemiological life are small case-control studies claiming to have found some evidence of a small, previously unsuspected link between some health risk and an implausible cause. The prototypical example of this is the supposed link between electromagnetic fields (EMFs) and childhood leukaemia, hints of which first emerged in the late 1970s. Over the years, the link has been examined in many case-control studies with hundreds of participants; when combined, these suggested a

Serious or spurious? Making sense of health headlines

Every observational study aspires to identify a genuine, causal connection between some health effect and some activity, from eating junk food to living near a nuclear reactor. Yet all they can really give us is more or less compelling evidence of some potential link. As the saying goes, 'correlation is not causation', and telling the two apart isn't straightforward. There are, however, some rules of thumb that can be used to decide which studies to take seriously, and which merit a 'Yeah, whatever'.

The most useful of these rules were suggested in the mid-1960s by Professor Sir Austin Bradford Hill of the University of London, whose observational study of smokers begun in the 1950s set a standard rarely matched since.[7] Inspired by Hill's criteria, the following is a handy checklist of what to look out for:

What kind of observational study is it? Is it a 'case-control' study? These typically struggle more with the problem of bias than 'prospective cohort' studies.

How surprising is the finding? Be especially sceptical of 'out of the blue' claims of previously unknown health effects – especially if the link is biologically implausible.

How big was the study? While 1,000 participants can seem large, by the time it's been sliced and diced to focus on certain groups, key findings can rest on *very* small numbers.

How big is the effect? If it's a surprising finding, many epidemiologists ignore anything less than a doubling in the benefit/risk from any single observational study. And if the inherent risk is small, then even doubling it may still not be worth worrying about.

How consistent is the link? Is there a convincing link between effect and exposure?

> **Where's the study been published?** Ignore claims made at conferences, and wait for publication in a respected journal. Even then, remember that publication is a necessary but not sufficient condition for being impressed. Top journals can and do print nonsense.

significant increase in leukaemia risk among children exposed to EMFs from appliances and power lines. Yet applying some epidemiological rules of thumb puts this disturbing conclusion in a different light. For instance, despite the apparent impressive numbers of participants in the studies, the most worrying risk increases came from those exposed to the highest EMFs – which typically involved just a few dozen cases and controls. In addition, exactly how EMFs should trigger leukaemia has never been plausibly explained – while there are plenty of potential sources of bias and misleading factors capable of mimicking such a link. All this suggests the evidence for a cancer risk from EMF is pretty feeble – and sure enough it has flip-flopped repeatedly. A 2007 review of the evidence by a team from the US Centers for Disease Control excluded EMFs from its list of significant environmental risk factors for leukaemia.[6]

At the top of the pecking order of observational studies sit the huge multi-centre prospective cohort studies able to control for many potentially misleading factors, resulting in compelling evidence of plausible risk factors. A classic example is the Million Women Study, set up in the mid-1990s by researchers at Oxford University. Focused on women aged at least 50, the study has looked for links between their health and a host of factors, from contraceptive use to diet and smoking. By the mid-2000s, the study had found evidence of a link between breast cancer risk and use of certain types of hormone replacement therapy (HRT). The link was both strong and plausible, and the sheer size of the cohort allowed the researchers to compensate for many potential biases without undermining the credibility of their findings.

It's entirely possible that over the decades to come, observational studies like the Million Woman Study will save millions of lives. They may not be as reliable as the gold standard of the blinded randomised controlled trial, but huge, well-managed prospective cohort studies are as good as it gets. On the other hand, the next time you read about some implausible health risk based on a small case-control study, relax, take a deep breath – and wait for it to be debunked.

↑UPSHOT

Observational studies can never be as reliable as the gold standard of the double-blind randomised controlled trial. But they're often the only way to cast light on critical questions. And if they're huge, well managed and their results not pushed too far, they too can be pretty trustworthy.

How a lot of bull sparked a revolution

No one knows exactly how the Great Pyramid of Giza was built, but you can bet it took longer and cost more than anyone expected. Over 4,500 years later, that's one thing that doesn't seem to have changed. From upgrading a computer system to building the International Space Station, no project is so high-flying that it can avoid being brought down to earth by unforeseen delays and cost overruns.

Which is odd, considering the effort poured into project management methods specifically designed to stop such debacles. With baffling names like 'Agile' and 'PRINCE2' and bizarre jargon ('Scrum of scrums', 'backlog grooming'), they certainly sound impressive. Yet it's not even clear they really work, whatever their advocates might claim.[1] Fortunately, research is now uncovering some pretty compelling evidence for the effectiveness of another means of foreseeing the unforeseeable. Ironically, it has its origins in a matter concerning what can truly be called a lot of bull.

The beast in question was a huge male ox, star attraction of the 1906 West of England Fat Stock and Poultry Exhibition in Plymouth, Devon. Attendees were invited to use their skill and judgement to estimate the beast's weight once it had been slaughtered. To make the challenge tougher, the organisers asked not for the live body weight, but for the so called 'dressed' weight – that is, the mass of the carcass, minus the head, feet, organs and hide. Around 800 people paid the current equivalent of around £5 to take part, and when the estimates were examined, one person had guessed the weight exactly, at 1,197 pounds (around 550 kilograms). But someone else won even bigger than day: the brilliant polymath Francis Galton. He decided to find out just how well people guesstimated the weight of the ox, and obtained all the entry cards for the competition. On analysing them, he made an extraordinary discovery. While their range was predictably pretty broad, their median (that is, the weight for which there were as many guesses below it as above it) turned out to be 1,208 pounds (555 kilograms) – within 1 per cent of the true weight.

How had the guesswork of all the individuals ended up producing a central value so close to the truth? Fluke was clearly one possibility, but writing up his findings in the journal *Nature*, Galton suggested a more intriguing explanation. He thought the competition had triggered a pooling of expertise. According to Galton, the imposition of the entrance fee had deterred many of the time-wasters and no-hopers, reducing what one might call 'stupidity bias'. Meanwhile, the prospect of winning encouraged skilled participants to do their best – boosting the accuracy still further. Combining the individual guesses thus gave a collective estimate based on the expertise of those willing to 'put their money where there mouth is'. And – in the case of the bull's weight, at least – the result was impressively accurate.

Now known as the 'Wisdom of Crowds effect', it has remained controversial ever since – not least because it seems to violate basic rules about extracting insight from limited information. Yet sceptics have had to face up to the mounting evidence for its effectiveness, such as the success of so-called prediction markets, which have powers that would have amazed even Galton. In the late 1980s, academics at the University of Iowa set up the Iowa Electronic Market (IEM), where pundits could buy and sell 'shares' in the outcome of US elections. The prices of the shares reflected the chances and margin of victory for each candidate. So, for example, if the share price implied an 80 per cent chance of a candidate winning, but someone thought the real chances were 85 per cent, the shares would seem a good bet, and worth buying. Those especially confident in their belief would be willing to buy lots of these shares, thus driving up the price – and hence the implied probability of victory. As participants focused on making money, their expertise ended up pooled, revealing the collective wisdom of the 'crowd' of pundits.

Over the decades, this crowd has proved astonishingly wise. A 2014 analysis by two researchers from the University of Iowa showed that the IEM beat conventional opinion poll results around three-quarters of the time, with a forecast error for candidate share in US presidential elections of just 1 per cent. The success of the IEM has since been mirrored by other prediction markets. Movie buffs can use their expertise to trade shares in the success of actors, new releases and Academy Award potential on the Hollywood Stock Exchange (HSX). Despite offering no more incentive than fake dollar fortunes and kudos, the predictions of HSX have proved so reliable that a spin-off has been set up to feed insights to Tinseltown executives. In one celebrated example, the collective wisdom of HSX spotted the hit potential of a $25,000 budget horror movie that studio management had ignored. It was called *The Blair Witch Project*, and grossed almost $250 million at the box office.

The wisdom of crowds can also be observed in so-called betting exchanges like Betfair. These match up punters holding opposing views, with one person's winnings coming from the lost bets of others, the exchange taking a small slice of the profit for doing the match-making. Punters are attracted by the fact that they generally get better odds than they would from bookmakers, whose higher running costs are reflected in less generous odds. Again, research has shown that the wisdom of crowds reflected in the final exchange odds is impressively reliable: outcomes whose chances of occurring are reckoned by the crowd to be worth, say, an evens-money bet really do occur around 50 per cent of the time.

As we'll see in later chapters, this increased accuracy of the odds actually makes it harder to succeed as a gambler. But it does show how the wisdom of crowds can produce reliable insights even in complex situations involving many interacting factors. And that has not gone unnoticed by those tasked with that age-old challenge of keeping projects on time and within budgets.

In the late 1990s, a team at the multinational technology company Siemens decided to find out whether the wisdom of crowds could do better than conventional project management in keeping a software project on track. Working with Gerhard Ortner of the University of Technology, Vienna, they set up a prediction market enabling those working on the project to buy and sell 'shares' whose price reflected the chances of the project meeting a deadline. The Siemens team set up two markets: one designed to flag up the risk of a delay, the other to capture insight into its likely length. The hope was that employees would give their insights quickly and anonymously via the market in order to bag a profit – thus giving earlier warning of trouble. And that's exactly what happened. The resulting market 'priced in' the impact of changes to the project long before their announcement by senior management, as employees raced to benefit from their personal insights and bought or sold shares. Within just a month of trading – and more than three months before the deadline itself – the markets were predicting the deadline would not be met, with the delay estimated at two to three weeks. With a month still to go, the markets were hit by a deluge of 'sell' orders, a clear sign that confidence in meeting the deadline had collapsed. Sure enough, the software project missed the deadline, and overran by two weeks. Meanwhile, the standard project management tools continued to insist all was well right up until the deadline.

Many companies have since experimented with 'wisdom of crowds' methods. Hewlett-Packard discovered that prediction markets gave more reliable forecasts of printer sales than its standard forecasting methods. Google found they helped predict future demand for products such as Gmail, and likely threats to its market share. An analysis of the performance of its internal prediction markets found an impressive correlation between the predicted chances of events according to the markets, and the frequency with which the events actually occurred. Ford, Procter & Gamble,

Lockheed Martin, Intel, General Electric – the list of corporations who have used prediction markets is a long one.

So if prediction markets are so wonderful, why isn't everyone using them all the time? The reasons are an intriguing mix of the rational and irrational. Sir Robert Worcester, founder of opinion pollsters MORI, doubtless spoke for many with his characterisation of prediction markets in 2001 as 'voodoo polls'. His concern focused on their apparent violation of the basic rules of sampling theory. First, prediction markets are anything but random samples; indeed, they're specifically designed to be biased towards including only those confident enough to risk their money or reputation. Secondly, prediction markets remain pretty reliable even when they involve just a few dozen 'traders' – a sample size that standard theory would deem dangerously small in many circumstances.

The mystery of how prediction markets can get away with such flagrant disregard for the rules has prompted much controversy and research – and some clues are now starting to emerge. One comes from the experiences of pollsters, who know all too well that a theory that works fine with coloured balls can't always be trusted with real live people. Over the years they've seen their apparently rigorous methodology trashed by people who tell them one thing, then do another. They've tried various tricks to correct for the effects of such dissembling, but to no obvious benefit.[2] This has led some researchers to wonder whether the wisdom of crowds effect is benefiting from its focus on the *traits* of the individuals making up the crowd.

This is a radical idea, akin to suggesting one can get good estimates of what's in a jar of coloured balls if they have a certain mix of colours. It also has implications for how best to arrive at collective decisions. But is it true? Research in fields as diverse as psychology, management studies, ecology and computer science has shown that when it comes to solving problems, there's certainly such a thing as having too much of a good thing. The problem isn't personality clashes or too many egos; it's simply that a high skill level often comes at the cost of narrowness. In 2004, Lu Hong and Scott Page of the University of Michigan proved mathematically that a group of moderately skilful people with diverse insights will typically solve problems more effectively than a team comprising solely the most skilful people.[3] This has obvious resonances with the wisdom of crowds effect, and the connection has been reinforced by a team led by decision theorist Clintin Davis-Stober of the University of Missouri.[4] They began by capturing the concept of crowd wisdom mathematically, and then looked at what can undermine it. Like Hong and Page, they found that the reliability of prediction markets depends on the traits of those taking part. Naturally, skill does play a role, but again diversity emerges as crucial. Once a prediction market includes some experts, the mathematics shows reliability is best improved *not* by recruiting more of the same, but by bringing in mavericks who think differently and/or have access to different sources of insight. Indeed, it's actually worth taking a hit on the skill levels of new recruits just to get more diversity. That's because the views of experts tend to be correlated, so bringing in more can turn small biases into major collective errors. The views of mavericks, in contrast, are by definition much less correlated with each other and everyone else. So while their biases may be larger, they're less likely to overwhelm the final collective view.

The work of Davis-Stober and his colleagues is part of a major ongoing effort to give the wisdom of crowds a solid theoretical basis. It has shown that collective wisdom can benefit from the insights even of rank amateurs, and remains robust even when some are trying to deliberately skew the outcome. In the process, the research has confirmed the value of including the insights of those who – as the corporate cliché has it – 'think outside of the box'. It has also cast light on why collective wisdom can emerge even from groups so small they barely merit the moniker 'crowd'. According to Iain Couzin of Princeton University and graduate student Albert Kao, the explanation again lies in correlation – this time between the sources of insight used by those in the crowd.[5] If these sources are widely available, they'll create correlations among those making a judgement – which is fine as long as the sources are reliable. But if they're not, the pooled judgement of a huge crowd is likely to be dominated by these correlations, leading to poor

How to tap the wisdom of your 'inner crowd'

While predictions based on collective beliefs can be impressively reliable, we don't actually need a crowd to benefit from its wisdom. We can do it all ourselves – if we're careful to include some crowd-like variety in our thinking. Stefan Herzog and Ralph Hertwig of the Max Planck Institute for Human Development have come up with a technique for doing this: dialectical bootstrapping.[6] Fortunately, it's simpler than it sounds. First, come up with an initial guesstimate of whatever you're trying to forecast using whatever insights you have, and note it. Now imagine you've been told it's wrong – and think about where you may have blundered. What assumptions could have been inaccurate, and what would be the impact of changing them? Would it make the resulting estimate higher or lower? Now make another estimate, based on your new view of the problem. Remarkably, Herzog and Hertwig have found that the average of the two guesstimates is typically closer to the true answer than either one individually.

reliability. In contrast, the average judgement of a small group is less precise, more diverse – and thus better protected against being undermined by faulty insights. There's an obvious limit to these benefits, however, and it's reached with a crowd of just one person. Ironically, the judgements of individuals have long been revered; indeed, their source often ends up with that hallowed title of 'guru'. That's not to say gurus must never be trusted; new research has identified methods allowing even those of us with no aspirations to guru status to make better judgements (see box above).

Many questions about the wisdom of crowds are still being investigated – for example, the optimal size of crowd for different judgement problems, the role of personality type and the benefits of giving feedback to participants. But one thing is clear: sceptics can no longer claim the evidence for the wisdom of crowds is purely anecdotal. There is now a substantial body of observational evidence, and it is increasingly underpinned by rigorous theory. Even so, the supposed lack of evidence and theory was probably never the real reason for the scepticism. Many people just have a visceral distrust of what to them is decision-making by mob. It's true that the rules governing the wisdom of crowds run counter to more familiar theory, and even common sense. Unlike samples of coloured balls from urns, smaller crowds are not necessarily less reliable than big ones. The 'obvious' importance of expertise is also more nuanced, with the addition of more mavericks likely to produce better collective wisdom than recruiting more 'authorities'.

Are we about to see a revolution in forecasting, where everything from construction projects to foreign policy is guided by the wisdom of crowds? Perhaps, but it's probably not worth asking your resident guru for a view.

↑UPSHOT

If faced with having to do some crystal-ball-gazing, be wary of trusting the confident claims of any individual – no matter how expert. Instead, set up a prediction market (perhaps via an online service like cultivatelabs.com) and invite everyone with a view to feed in their insights in return for cash or kudos. Research suggests the resulting collective wisdom will prove much more reliable than that of any supposed 'guru'.

How to beat casinos at their own game

One Friday night in August 2014, Walter and Linda Misco from New Hampshire walked into the MGM Grand casino in Las Vegas, and headed straight for the bright, shiny loser-magnets known as the slot machines. Since their invention over a century ago, these 'one-armed bandits' have swapped their eponymous pull-lever for buttons and electronics, but they've lost none of their ability to relieve people of money. This didn't faze the Miscos; indeed, they wanted to find the most notorious slot machine in Sin City: the Grand's Lion's Share machine. One of the casino's original slot machines, it had acquired worldwide infamy for having never paid out a single jackpot since it was installed in 1993. There was a flip-side to this notoriety, however: as a so-called 'progressive' machine, the Lion's Share's stinginess meant the jackpot on offer had grown to the point where the winner would become an instant millionaire. And over the years, the machine attracted players from around the world, who cheerfully stood in line to have their shot.

When the Miscos got their chance, they fed in $100 to bankroll their bets, more in hope than expectation. But just five minutes into their session, three green MGM lion heads all appeared in a line. The lights flashed, the machine blared and then it dawned on the Miscos that they had done what no one had done before, and won the Lion's Share: all $2.4 million of it.

For many, this is one of those heart-warming stories of Lady Luck finally doing the decent thing. Certainly that's how the media saw it, and the Miscos obliged by revealing that they planned to use the money to put their grandchildren through college, and buy a sports car. But to others, what happened to the Miscos simply highlights all that is wrong with casinos, and their cynical ploys to keep mug punters coming through the doors.

Everyone has a view on casinos. Some are entranced by the glitzy, glamorous image portrayed in movies like *Ocean's Eleven* and *Casino Royale*. Others are repelled by the thought of deadbeats feeding their life savings into machines. Yet for anyone who wants to really understand chance, a visit to a casino is a must. They are temples to probabilistic cunning. With revenues in excess of $150 billion a year, the world's casinos provide compelling proof of the benefits of having a branch of mathematics at the core of a business model – especially one most people think they understand, but don't. Their image may be tainted by associations with people keener to use fists than brains, but casinos owe their success to clever use of that most misunderstood of probabilistic theorems, the Law of Averages. Most of the well-known games they offer, including roulette, craps and slot machines, have outcomes whose probabilities can be calculated precisely from first principles. And armed with these, casinos have created a business model based on payouts which look reasonable, but aren't. They're all less than they should be for a genuinely fair game – but, cunningly, most of them are not egregiously unfair either. It's a combination that pulls off the remarkable trick of ensuring lots of punters come out on top, while 'the house' still gets a rock-solid profit margin.

Take the quintessential casino game of roulette, with its famous wheel of 36 alternating red and black number 'pockets'. As there are 18 of each colour, it seems obvious that the probability of the ball landing in red or black is 50:50; certainly that's what the casinos want you to think, as they pay out at evens odds to anyone who bets on red or black. But take another look at the wheel: tucked in unobtrusively among the red and black pockets is another, numbered zero and coloured green; in the USA there's usually a second green pocket, numbered '00'. It hardly seems important, and one can easily sit through dozens of spins without the ball landing on green. But a quick sum reveals something odd. Suppose you're in a Las Vegas casino, and you bet on red. The chances of bagging the evens odds on offer are given by the number of red pockets – 18 – divided by the total number of pockets into which the ball could land, which is 38, as we have to include the two green pockets as possibilities. So the chances of winning the evens bet are not 18 divided by 36, but 18 divided by 38 – which is 47.37 per cent, rather than 50 per cent.

That seems unfair, and it is. Those green pockets have tilted the game towards the casino. But here's the thing: the tilt is so slight – less than 3 per cent – that it's easily swamped by the random fluctuations in the short term … such as the time spent at the table by most punters. Over the course of a few hours, some may win big, others will curse their luck – but none would be able to detect the small bias in favour of the house. Indeed, the Law of Averages shows it would show up convincingly only after careful observation over at least 1,000 spins. And who plays for that long? The casinos, that's who – via dozens of wheels, 24 hours a day, 365 days a year. So while any individual punter won't feel cheated, the Law of Averages ensures their collective efforts give the casino a pretty much rock-solid profit margin or 'house edge' of 2/38th or 5.3 per cent from all the red/black bets (or 2.7 per cent in European casinos).

So is it possible to beat the casinos? Over the years, many people have tried their luck with various simple strategies, only to find out that their luck runs out. Anyone familiar with the Law of Averages knows that tricks like betting on 'runs' of reds to end won't work: the ball retains no memory of what it's done before, and so the odds remain unchanged for each spin. Casinos are happy to talk up the supposed benefits of making bets according to a 'staking method' like the martingale – basically double or quits – or more exotic ones like Labouchère staking or the d'Alembert method (the fact that the eponymous eighteenth-century mathematician failed to understand coin-tosses tells you all you need to know about this one). They all claim to combat the vagaries of chance by upping the size of bets when conditions are 'favourable' and cutting back when they're not. They may turn a profit for a while, but in the end they all fail for the same reasons. First, the casinos won't allow you to keep upping your bets according to some staking strategy; they all impose a 'house limit' to manage their risk exposure. And then there's the Law of Averages, which ensures that as you continue to play you'll increasingly feel the depredations of the house edge, no matter how small it is. The combination prevents any 'staking plan' from turning an unfair game into a reliable revenue stream.

Even so, there are ways of making money in casinos – which don't involve cheating. They rely on loopholes in the seemingly unimpeachable Law of Averages. Recall that the law states that the probability of an event resulting from some random process can be estimated ever more precisely by dividing the number of times it occurs by the ever-increasing number of opportunities it is given. So, for example, in roulette, the proportion of times the ball lands in red will get ever closer to the theoretical value of 47.37 per cent as the number of spins increases.

But lurking in that mathematically unimpeachable result are several caveats. The most obvious is the assumption that the process propelling the game really is random. As we saw with the toss of a coin (Chapter 1), what seems random and unpredictable can actually just be extremely complicated and at least broadly predictable. In the case of roulette, the bouncing ball is ultimately subject to the laws of physics, and as such its motion cannot be genuinely random, which by definition means obeying no rules whatsoever.

This loophole in the Law of Averages has underpinned many successful attempts at extracting money from casinos. As an engineer working in the Victorian cotton industry, Joseph Jagger knew that mechanical devices don't always work exactly as intended. This led him to wonder whether there were flaws in the operation of roulette wheels which might be exploitable. In 1873, he sent a team to Monte Carlo to surreptitiously monitor the performance of the roulette wheels at the Beaux Arts Casino. Sure enough, they found that the balls were more likely to land in some sectors of the wheel than others. The bias was too small to be spotted by the management, but – crucially – it was big enough to overcome the wafer-thin house edge on some of the roulette bets. This turned the slightly unfair odds into profitable ones for certain bets on certain numbers. Armed with this insight, Jagger went to Monte Carlo and over a few days in July 1875 won the equivalent of around £3 million today – making him the best real-life claimant to the title of the Man Who Broke the Bank at Monte Carlo.

Casinos have long since realised the critical importance of regularly checking all their gaming equipment for flaws, wear and tear and malfeasance. But that doesn't entirely close the loophole, as even brand-new, perfectly adjusted roulette wheels are subject to laws of physics that offer at least some predictability. In 1961 mathematicians Claude Shannon and Ed Thorp – arguably the two finest minds ever to take on casinos – built a computer capable of turning observations of how and where a roulette ball was set running into predictions of the four or five numbers where it would land. This transformed the slim profit margin of the casino into a hefty 40 cent edge for Shannon and Thorp. Technical problems prevented the pair taking it to a casino, but the idea was revived in the late 1970s by a team of physics students from the University of Santa Cruz. They fitted a microprocessor capable of doing the sums into a cowboy boot, headed to Vegas, and allegedly made a tidy profit.

This strategy of using the laws of physics to prise open loopholes in the Law of Averages is now being combined with ever more sophisticated technology. In March 2004 a trio of east Europeans took over £1.3 million at the Ritz casino in London using a laser hidden in a fake mobile phone to gather the data needed to predict where the ball would land. After analysing video footage, the casino called in the police, but the trio were released without charge and allowed to keep their winnings.

Another, more subtle, loophole in the Law of Averages allows money to be made by playing one of the most popular card games in casinos: blackjack, or '21'. Put simply, this involves players and the dealer both receiving cards, and the players betting they can reach a total card value closer to 21 – or even exactly 21 ('blackjack') – than the dealer. While rules vary, they typically lead to the game being unfair, though the house edge is wafer thin, at less than 1 per cent. But there's a loophole buried in the calculation of this edge that skilful players can exploit. The cards are dealt from a shuffled set of decks – typically around half a dozen – and then discarded rather than added back to the pack (a process mathematicians call 'sampling without replacement'). As such, while specific card values may appear at random, there's not an endless supply of them: with, say, four packs in the 'shoe', once you've seen sixteen aces, you're not going to see another until the shoe is reshuffled. And that means the chances of getting winning blackjack hands – unlike those in other casino games such as roulette – aren't fixed, but change as play proceeds. This loosens the grip of the Law of Averages, and allows the chances of winning to turn significantly in favour of players. Better still, it also undermines the rule that there's no way to turn an unfair game into a profitable one simply by betting in a particular way. In blackjack, you hold back while the chances of winning are against you, and pile in when they go in your favour.

Spotting when that happens involves the technique known as 'card counting'. Devised by mathematician Ed Thorp, who publicised the technique in his 1962 best-seller *Beat the Dealer* (still in print to this day), card-counting was initially dismissed by casinos as just another 'get rich quick' scheme. But the truth was they'd blundered: they had thought shuffling the packs was enough to lock in their house edge. They'd missed the fact that the act of playing reveals the identities of cards emerging from the shoe, and thus gives insight into what is likely to happen next.

Thorp devised a system for keeping track of what cards had already appeared and amending one's betting accordingly. As the impact of card-counting is relatively small, it demands a hefty bankroll and sustained concentration to turn it into a decent profit. Even so, publication of Thorp's book led to casinos being taken for substantial amounts by everyone from college students to retirees who took the trouble to master card-counting. So the casinos fought back. They began by simply increasing the number of packs used to at least half a dozen – increasing the mental demand of card-counting. Then they brought in automatic shufflers which re-randomised the cards halfway through a session, trashing ongoing counts. Their speed also boosted the number of deals per hour, thus giving the house edge more time to work its magic. Many casinos have simply changed the standard payout odds for blackjack, thus cancelling out the tiny advantage of card-counting.

Despite all this, there are still plenty of card counters out there, and for them casinos reserve the ultimate countermeasure: the Quiet Word. While not illegal, even suspected card-counting is deemed unacceptable by most casinos – and they don't care who knows it. In 2014 the Hollywood star (and accomplished card player) Ben Affleck was reported to have received a Quiet Word from the management at the Hard Rock Casino that he was welcome to play any other game – Vegas-speak for 'We think you're card-counting and you're going to stop'.

Arguably the single most effective strategy for making a fortune in a casino is to make clear you already have one – thus making you a 'whale' in Vegas-speak. Casinos love whales as they spend big, lose bigger but can cover their debts. As a result, casinos cheerfully accommodate pretty much every whim a whale might have. That's what gambling magnate and blackjack specialist Donald Johnson was hoping would happen to him when he hit several casinos in Atlantic City in 2011 with spectacular results. Having made clear he would be playing for $25,000 a hand, Johnson succeeded in negotiating a host of tweaks to the standard rules of blackjack, all of which reduced the house edge. Then he unleashed the two strategies normally associated with the old Mafia-run casinos: distraction and intimidation. Johnson turned up to every game with a party of provocatively dressed women. Their presence plus the unnerving size of the bets he made led to the casino dealers losing concentration and making mistakes. That compelled them to give Johnson free bets, which finally tipped the edge his way.

Over the course of several months, Johnson took his strategy to various casinos in Atlantic City and relieved them of around $15 million. His coup led to headlines, managers being fired – and the inevitable Quiet Word from many casinos telling him he was no longer welcome.

So the age-old claim is true: there really are ways of beating casinos at their own game. The bad news is that it involves precisely that: having comparable levels of skill, determination and money. But most people who go to casinos aren't planning to make a career out of it; they go for the fun of it, plus the allure of perhaps winning a bit of money. And the good news is that – as we'll see in the next chapter – the laws of probability lead to some top tips for maximising the chances of doing both.

↑UPSHOT

Casinos are factories that use the laws of probability to produce profit. Loopholes in those laws do make it possible to divert some of that profit your way, but they're small and squeezing through them demands skill, determination and a lot of money.

Where wise-guys go wrong

When the Miscos walked out of the MGM Grand in Las Vegas with $2.4 million they never claimed to have been anything other than lucky. They just happened to be playing the Lion's Share slot machine on the day it paid out its first jackpot in 21 years. Over that time, the machine had made the casino over $10 million in profit in a revenue stream underpinned by the Law of Averages. Clearly, whatever else the Miscos got, it sure wasn't the lion's share of what the slot machine had taken. No serious gambler would go anywhere near slot machines, with their hefty 5 to 15 per cent house edge and zero scope for skill. Instead, they focus on the low-edge games like blackjack and baccarat, planning to use their skill in strategies like card-counting to turn a profit.

Yet even the smartest players can fall into the trap of thinking they're winning because of their skill, while all the time the Law of Averages has just been biding its time. And if they're playing a game with an ineluctable house edge, such as baccarat, the law will catch up with them in the end. Knowing when to quit is thus a key skill of any professional gambler. Even so, it can elude even the smartest players.

As a Japanese real estate tycoon, Akio Kashiwagi was both smart and rich. He was also addicted to baccarat, with a relentless style of play that earned him the title 'The Warrior'. He thought nothing of taking on casinos in $100,000-a-pop games which went on for days. Among casino managers Kashiwagi was a 'whale', wealthy, confident and willing to bet big. As a result, he was 'comped' on a lavish scale: food, drink, VIP rooms, even flights to and from the casino were all supplied compliments of the management. The aim of the casino was simple: to keep him there long enough for the Law of Averages to destroy him. Unlike blackjack, baccarat is a game whose house edge cannot be defeated by skilful play or betting. In 1957, two mathematicians had found an optimal way to play baccarat, but all this did was defer the inevitable. Play the game long enough, and the Law of Averages will get you.

And so it was that in May 1990 Kashiwagi sat down to a baccarat session arranged especially for him by the newly opened Trump Taj Mahal casino in Atlantic City. The stakes were $200,000 a hand, and the game was to continue until either he or the casino won $12 million. Kashiwagi lived up to his reputation, playing with skill and tenacity, and amassing $10 million. But then the wafer-thin edge started to turn against him, and he made the classic mistake of all mug punters: he began chasing his losses. As the hours went by, Kashiwagi's losses mounted. Finally, after 70 hours of play over six days, he grabbed $2 million in chips and quit.

But then the Taj's strategy started to unravel. It had gambled that Kashiwagi was good for the $10 million he owed. But the casino was still short of $6 million in January 1992, when Kashiwagi was found dead in his home near Mount Fuji. He had been stabbed over a hundred times – some believe on the orders of the Yakuza, Japan's equivalent of the Mafia. Bizarrely, he went on to achieve a kind of immortality via a scene in Martin Scorsese's 1995 film *Casino*. Some of the details were changed – Kashiwagi in Atlantic City became 'K. K. Ichikawa' in Vegas – but the outcome and the moral were the same. He starts by winning but gets greedy, plays a game like baccarat too long – and suffers the consequences. The words of the fictional casino manager Sam Rothstein (based on real-life casino boss Frank 'Lefty' Rosenthal) make clear the strategy used by the casino to reel in a whale: 'The cardinal rule is to keep 'em playing and keep 'em coming back. The longer they play, the more they lose. In the end, we get it all.'

Yet casinos need more than whales to succeed, and even the

heftiest house edge is worthless if there are no punters coming through the doors. And that basic truth supplied a final twist to the tale of the baccarat-playing whale. In 2014 five of the biggest casinos in Atlantic City closed for lack of business; they included Kashiwagi's nemesis, the Taj Mahal.

Most people who go to casinos aren't whales, but they are still at risk of being reeled in if they venture out of their depth. It's vital to know how to spot the danger signs, how to get the most off the 'hooks' dangled by the casinos, and what tempting bait to avoid altogether. That means applying the laws of probability. While the mathematics behind them is surprisingly complex and still provokes controversy, they're easy enough to apply, and make intuitive sense.

The first exploits the fact that randomness usually takes time to reveal itself. Toss a coin a few times, and it's perfectly possible to get all heads or all tails, suggesting the coin isn't behaving randomly at all. Keep going, though, and the fact that there are two possible outcomes will become increasingly clear. This is symptomatic of what mathematicians call the 'asymptotic' nature of the Law of Averages – that is, what it says about relative frequencies only strictly applies to an infinitely large sequence of events. For any finite sequence, a whole range of possibilities is consistent with randomness, and it can be radically different from the long-term average for really short sequences.

Applied to casino games, this means that during short sessions one can get some pretty hefty departures from the house edge, or profit margin – and if that house edge was pretty thin to start with, the result is a burst of profitability for players. The shortest of short sessions is of course a single play. While even this won't turn the odds in your favour, it does minimise the time you're exposed to the Law of Averages – and thus the time during which the house edge can make itself felt. The single-play strategy was adopted by Ashley Revell in the spectacular $135,000 winning bet described in the Introduction. He was smart in just playing once – but also lucky.

Such bold play is not for the faint hearted, nor is it much good for those keen to experience the atmosphere of a casino. So a compromise is needed, and the best is to seek out games with the smallest house edge and play long enough to stand a good chance of coming out ahead, but not so long that the Law of Averages starts to make itself felt.

To achieve the first goal, avoid slot machines and lottery games like keno, whose eye-popping jackpots are funded by eye-popping house edges. Instead, focus on simple bets on roulette (such as red/black), or learn how to play and exploit the low-edge bets in games like blackjack and craps. Next, decide how much time and money you're prepared to spend in the casino, and play until one or the other has run out. But don't spend your time making lots of little bets, as that reduces your chances of coming out ahead. For example, suppose you go into a casino with £100 and decide to try your luck at roulette. Depending on how busy it is, you'll get around 30–40 spins per hour. You've got a better chance of at least breaking even if you spend fifteen minutes making £10 bets than taking half an hour to place £5 bets. That's because you'll make just ten bets in the first case, and 20 bets in the second – and by halving your exposure to the house edge, you'll boost your chances of making £50 profit before going bust from barely 1 in 3 to around evens. If you make around ten red/black bets, mathematically you'll have a better than evens chance of at least breaking even, almost a 1 in 3 chance of making a profit – and a 100 per cent chance of being able to say you played roulette and knew what you were doing.

Don't be too ambitious in your goals, either, such as aiming to stay until you've doubled your bankroll. More modest aims have more chance of being achieved. So, for example, while you've got an evens chance of turning £100 into £150 before going bust playing red/black with £10 bets, those chances are cut in half if you aim for £200. And don't of course fall for any baloney about 'riding your luck' if you reach your target early. Take your money and run – before the beast of randomness wakes and devours it all.

Follow these rules,[1] and you'll have a better chance of emerging from your visit to a temple to probabilistic cunning with a smile on your face.

Even professional gamblers can mistake luck for skill and spend too long in the deadly embrace of the Law of Large Numbers trying to boost profit or win back losses. The trick to having a good time in casinos is to turn up the discipline, turn down the ambition, and cut your losses.

The Golden Rule of Gambling

I f casinos represent the glamorous side of gambling, backstreet bookmakers are their antimatter equivalent. Tawdry, joyless and faintly menacing, they have long been notorious as the haunts of deadbeats and desperados. Yet they also bear witness to the popularity of a form of gambling that dwarfs the likes of roulette and blackjack. This is sports betting: gambling on the outcome of everything from which horse will win the Grand National to how many yellow cards will be dished out during a football match.

Sports betting is a colossal global enterprise, generating revenues estimated at around a trillion dollars a year. In the city of Hong Kong, the sport of horse racing alone generates turnover of over $10 billion. Betting on single sporting events, such as the Superbowl, has reached similar levels.

Such staggering sums speak to the fact that hundreds of millions of us enjoy the occasional 'flutter', backing our beliefs with hard cash. And with the advent of internet betting, it's never been easier to do so. Bet365, Britain's biggest online bookmaker, saw over £34 billion flow through its books in 2015. Despite long being frowned upon in polite society, no amount of finger-wagging has been able to stop the increasing popularity of gambling. Yet those who regard being able to back a horse at Epsom as an inalienable right must face the fact that most regular sports bettors lose money – sometimes with disastrous effect. And while it's tempting to blame it all on the bookmakers, the real reason is as clear as day:

most regular punters don't really understand what they're doing. They may know how to read a form book and how to place an each-way bet, but they've no idea how to spot a decent bet and turn a profit. Which is, of course, exactly how the bookmakers generate theirs.

An estimated 95 per cent of those who bet on sports fail to turn a consistent profit.[1] So what is it that the other 5 per cent know that everyone else doesn't? Surprisingly enough, nothing very complicated; indeed, the wonder is that so few are aware of it. The challenge lies in putting it into practice. Only those who have actually tried it can understand how so simple a principle can play havoc with your sanity.

Few have mastered the art and science of gambling as successfully as the English horse racing punter Patrick Veitch.[2] He has become a multimillionaire from his exploits, along with the title Enemy Number One among Britain's bookmakers. But the background to his success should serve warning on anyone dreaming of emulating him.

Veitch is, first and foremost, extremely smart. At the age of just fifteen, he won a place at Trinity College, Cambridge – Isaac Newton's alma mater – to read mathematics. Barred from starting his studies because of his youth, he spent the late 1980s honing his skills as a serious gambler. He quickly focused on horse racing, attracted not by its image or popularity but by its *complexity*. A runner's chances of winning depend on a host of factors, from its past performances and the quality of its rivals to the shape and 'going' of the racetrack on the day. Quite apart from the intellectual challenge, the teenaged Veitch had already identified something that forever eludes most gamblers: that the complexity gave him the best chance of spotting factors overlooked by everyone else – including the bookmakers when drawing up their 'tissue' of odds for each race. This was an early sign of the gambling strategy that became the basis for Veitch's fortune.

Once at Trinity, Veitch quickly distinguished himself in applied mathematics, though not quite the type being studied by the other students. While they sat through lectures on vector calculus, he went to race meetings, routinely putting on bets of £1,000. He then started up a tipping service; it proved so successful that by the start of his final year Veitch had £10,000 a month flowing through his books. Deciding he was wasting his time studying mathematics, Veitch dropped out of Cambridge before taking his degree.

Chances are Veitch never went to any undergraduate lectures on probability. Had he done so, he would have encountered the standard proofs of the Law of Averages (or the Weak Law of Large Numbers, as it is unhelpfully named by academics) and learned its implication: that in the long run, the true probability of any chance event is revealed ever more precisely by the number of times the event occurs, divided by the number of opportunities it has to do so. Doubtless the students were given problem sheets to solve with exercises about the probability of certain outcomes of coin-tosses and dice-throws. Yet all this would have been of little interest or use to Veitch, because it focused on the wrong type of probability.

The idea that there are different types of probability has sparked bitter debates among academics for centuries. We'll encounter some of the unhappy consequences of this controversy in later chapters. It has generated a lot of jargon ('aleatory' versus 'epistemic' probability, frequentism versus Bayesianism), philosophical musing and mathematics. But the basic notion of different forms of probability is easy to grasp, via the difference between casinos and bookmakers. Casinos know the chances of all the various outcomes in games like roulette, craps and slot machines. The probabilities don't have to be guessed at or estimated from raw data. They can be found from first principles. There are 38 pockets where a Vegas roulette ball can land, so the chances of landing on any one of them is 1 in 38, and it's just as likely to land in one as in any other. So that's the random (in the jargon, 'aleatory' – pronounced ale-a-tree) probability of that event, and it allows the casinos to know that the Law of Averages will work its magic for them. Bookmakers, in contrast, have no such assurances. That's because it's just not possible to work out the probability of, say, a horse winning a race from first principles. Unlike the spin of a roulette wheel, the outcome of a race depends on a complex mix of variables, from the

An odds way of speaking

Sports gamblers are in it for the money (at least, in theory) and so traditionally describe the chances of events not as probabilities, but as the profit generated by a winning bet. So, for example, instead of saying a horse has a 22 per cent chance of winning, they'll say it's '7 to 2', meaning that for every £2 bet, the fair profit for a win would be £7. To convert odds of 'X to Y' to a percentage probability, divide Y by X + Y and multiply by 100. For high-probability events, gamblers talk of an event being '3 to 1 on', meaning just £1 profit for every £3 bet. To convert these to percentages, just swap the Xs and Ys in the formula – so, for example, '3 to 1 on' becomes 75 per cent.

fitness of the horse to the jockey to the state of the racecourse. As such, bookmakers must rely on their own judgement of a horse's chances (its 'epistemic probability', in the jargon), and use this to set their odds.

What bookmakers do share with casinos, however, is a determination to make a profit by offering payouts somewhat less generous than they should be. To see how this works, suppose a bookmaker's odds-setters believe that a horse has a 40 per cent chance of winning ('6 to 4' in bookie-speak, meaning a win gives £6 profit for every £4 bet – see box above). The actual odds posted by the bookmaker won't be 6 to 4, though, but closer to 'evens', implying a 50 per cent chance of winning. As their name implies, evens odds give £4 profit for every £4 bet – which is much less generous than the 6 to 4 odds would pay out. In other words, the payout being offered is unfair to the punter, and the bookmaker pockets the difference as profit. Anyone who thinks bookmakers' odds accurately reflect the chances of the event happening has thus fallen straight into a trap.

Their posted odds are the equivalent of the casino trick of appearing to offer a fair payout while actually doing nothing of the sort – with the difference being the profit margin (sometimes called the 'overround' or 'vigorish'), often 20 per cent or more.

That may sound a pretty lucrative business model, but it's far less reliable than the house edge of casinos. That's because the odds are based on judgement – and a horse race, or indeed any sporting event, can fail to follow the script. Bookmakers try to cover themselves against this by offering unfair odds for every possible outcome of the event – say, a home win, away win or draw for a football match. While they have to strike a balance with what competitors are offering and punters will accept, their aim is to give themselves a 'balanced book' with a good chance of giving a decent profit margin regardless of the outcome.

Take the real-life case of the odds offered by one bookmaker on a Euro 2016 qualifying football match, with the odds converted to the probabilities they imply for each outcome:

Outcome	England win	Slovenia win	Both teams draw
Odds	11 to 4 on	10 to 1	4 to 1
Corresponding probability	73 per cent	9 per cent	20 per cent

This all seems to make sense: there are only three outcomes to the match – a win by either side, or a draw – all have been given odds, and England is given a better chance of winning than Slovenia, though a draw is certainly possible. But look more closely, and the effect of the bookmaker's determination to make a profit becomes clear. As one of the three outcomes must happen, their individual chances must add up to 100 per cent. Yet the total in the above 'book' comes to 102 per cent. That's the telltale sign that these offered odds aren't the bookmaker's actual belief about the chances of each outcome, as these would have to add up to 100 per cent. In other words, the real chances for at least one of the

The Golden Rule of Gambling

Making regular money from betting demands a proven method for identifying 'value bets'. These require that the true chances of the event occurring are significantly *higher* than the bookmaker's odds suggest. Identifying value bets thus demands the discovery of factors affecting outcomes that even bookmakers have not fully taken into account when estimating the true chances of the event. Without a proven method of finding and exploiting such factors, betting will eventually lead to substantial losses.

outcomes in the bookmaker's view are lower than those quoted – and the 2 per cent difference is being pocketed as profit.

In fairness, bookmakers need considerable skill in order to set even these unfair odds, as only if they're based on an accurate estimate of the true odds will they become a source of profit. If odd-setters get their estimates of the true odds wrong, they'll end up inadvertently offering overly generous odds. And that's where the likes of Veitch and other successful gamblers come in. They use their own skill to estimate the real odds of each outcome, and then compare them to what the bookies are offering. Their aim is to find so-called 'value' bets – cases where the bookies have missed something crucial in their analysis, and have thus offered overly generous odds.

What they're doing requires huge skill and determination, but its essence can be summed up in a simple formula, which one might call the Golden Rule of Gambling (see box above).

The Golden Rule simply crystallises the fact that the odds offered by bookmakers cannot be taken at face value. They've been deliberately concocted to pay out *less* than they really should, in light of the bookmakers' estimate of the true probability of the event, which is usually considerably lower than the odds imply. As such, anyone who relies on the bookmakers' odds to gauge the chances of a win will eventually amass a hefty loss.

Of course, if you're only putting occasional bets on big events for a bit of fun, none of this matters very much. The difference between the true and posted odds is usually small enough that they can at least be taken as a rough guide to the relative rankings of the various outcomes. Short-odds favourites really do tend to win more often than rank outsiders. But the danger comes if you make so many fun bets that the difference starts to reveal itself, as the long-term effect of the Law of Averages kicks in.

So, for example, someone who bet on the favourite in every one of the 144,000 races that took place in the UK in the 20 years up to 2010 would have seen their horse win in around 1 in 3 races. That sounds pretty impressive, and it would lead to an equally impressive profit if the odds for favourites were significantly more than 2 to 1. But they're not: the bookmakers typically offer worse than evens odds for favourites. As a result, while around a third of all your bets may win, the losses generated by the other two-thirds will eat up all your winnings, and more. The record shows, in fact, that if you had put £10 on every favourite over those 20 years, you'd have ended up with a net loss of well over £100,000.

In contrast, the Golden Rule of Gambling shows that there is a way to make money as a regular gambler. Like casino gambling it involves a skill, and in this case it is spotting where the bookmakers have blundered and offered better odds than they should. But they can't just be only slightly better: the size of the blunder has to be big enough to include some margin of error in your own judgement, plus a profit margin. So, for example, imagine you think a runner in the 2.30 at Ascot has a decent chance of winning, and that the bookies are offering odds of 3 to 1. The Golden Rule says you should put on a bet solely if you're confident not only that the runner has a 'decent chance', but specifically that it has a probability significantly greater than that implied by the bookmaker's odds, namely 25 per cent. Add in a margin of safety, plus a profit margin, and the Golden Rule is telling you that betting on this runner makes sense only if its chances of winning are at least 35 per cent. Are you really confident the bookmakers have blundered to that degree?

This is the question that trips up most aspiring professional gamblers. They believe the key question is simply who will win. In their search for the answer, they may put in hours a day studying 'form books', specialist publications and online sites to build up a really detailed picture of, say, some football team or tennis player – and spot when they're in with a real chance. The team's star striker is back from injury, say, or the tennis player is doing well on clay courts. Armed with these insights, they then put a bet on with a bookmaker. What they've overlooked is that the bookmakers' experts have access to all the same information and far more besides, and have then done their best to offer unfair odds on every possible winner. As such, the profits made by each win will simply not compensate for all the losing bets – thus ensuring that the gambler makes a long-term loss.

For gamblers who fail to ask the right question, the occasional – even frequent – wins are almost a liability, as they help mask the long-term consequences. Only as the weeks, months and years go by does it become clear that the wins aren't turning into regular profits. The Law of Averages is slowly but steadily destroying them.

Successful sports bettors like Veitch achieve their radically different outcomes by taking a radically different approach. Their focus is not on spotting winners but on finding outcomes whose chances have been significantly underrated by bookmakers. This can lead them to act in ways that baffle amateur gamblers, such as betting on several horses in a race. If your focus is on finding winners, this makes no sense as there can only be one. But for those who know that finding value bets is the key, then it is entirely possible to find several examples in a single race.

Doing so, however, is quite a different matter – and some suspect it is now all but impossible. Once upon a time, there were plenty of opportunities for determined sports bettors to make money. While bookmakers focused their attention on the major leagues of popular sports, specialist gamblers could scour the odds posted for games in lower leagues or in minority sports, and find misplaced bets. While the rewards were pretty modest, it was still hard work. But since the mid-2000s, it's not clear that any amount of hard work can make fortunes in sports gambling. All the big-name bookmakers now base their odds on sophisticated statistical analysis of past data, combined with computer modelling by specialist consultancies. In addition, they make extensive use of the odds produced by betting exchanges like Betfair, which are based on the insights of thousands of individuals, and the colossal Asian betting markets. The resulting odds are the product of the 'wisdom of the crowd', and are exceptionally reliable. That is, over thousands of sporting events, those with outcomes carrying odds of, say, 3 to 1 really do come to pass around 25 per cent of the time. As the betting exchanges make money via a radically different business model to bookmakers' (namely, by taking a percentage off winning bets), their odds really are estimates of true probabilities, rather than misleading odds deliberately downgraded to give a profit margin. All this means that it's never been harder to find blunders by bookmakers, and thus to place value bets. As an economist would put it, the sports betting market has never been more 'efficient', with posted odds reflecting essentially all the information available to anyone. The advent of so-called betting 'bots' – computer algorithms that detect any misplaced odds on betting exchanges – has led to even temporary inefficiencies vanishing in seconds.

Even so, there may still be scope for making a little money from sports betting for those prepared to put in the effort. The trick is to analyse past data, looking for factors that bookmakers have overlooked, creating inefficiencies and thus value bets with overly generous odds. One such factor is the number of runners in horse races. A large 'field' is more challenging for bookmakers to price up accurately, leading to high-odds outsiders being missed – but it can also lead to no-hopers getting in the way of better runners. On the other hand, small fields are easier to assess, and offer less opportunity for surprises. So somewhere in between – say, races involving between half a dozen to ten runners – there's potentially opportunity for spotting value bets. Another route is to focus on 'novelty' markets, such as how many shots a team has on target. Bookmakers put relatively little effort into analysing these, and so may overlook factors that lead to value bets. Whatever path is chosen, finding and validating such factors involves 'data mining', and as we'll see later, this holds many traps for the unwary. Certainly, those who succeed don't brag about how they did it. That's because once such factors become widely known, they'll end up being 'priced in' to the published odds – thus destroying any value they might contain. As Nick Mordin, a British horse-racing system analyst and tipster puts it: 'Betting systems are like vampires: when you drag them out into daylight, they die.'

Is there an easier way to making money out of gambling? Yes, if the claims made on countless websites are to be believed. They advertise books, computer software and tipping services that claim to be able to identify winners. Do they work? Many do indeed spot a good number of winners, but as the Golden Rule of Gambling shows, that's neither especially difficult, nor the point. The only (legal) way to make a long-term profit from gambling is by identifying value bets. Some of the tipping services can reasonably claim to do this, but the problem then is one of greed. Once a tipping service proves to be reliable, it will inevitably attract those with more money than sense, who try to put massive amounts of money on with the bookmakers. Always alert to new threats, the bookmakers will respond by dropping the odds to protect their margin – and thus destroy any value. And that's if the bookmakers will actually accept your money. For serious professional gamblers like Veitch, spotting value bets is one thing; being able to exploit them with serious money is another matter. Bookmakers keen to protect their cashflow can and do refuse to take bets from those they think really know what they're doing, making 'getting on' a major challenge. Online bookmakers run software that flags up punters whose success threatens their business models, and 'stake limit' them to ludicrously low bets, or simply close their accounts.

Most people who 'have a flutter' do so only for fun – perhaps once a year on a big event like the Grand National horse race in the UK or the Superbowl or Kentucky Derby in the USA. They never think of it as a way of making a living. Which is just as well, as most people are unaware of the Golden Rule of Gambling, let alone its implications for successful gambling. The reality is that, as with casino gambling, unless one is willing to put in a lot of effort, the most likely way of making a small fortune through gambling is to start with a large one.

↑UPSHOT

It's entirely possible to be a successful gambler. It just requires three things: an understanding of the Golden Rule of Gambling, expertise that can find opportunities that comply with it – and the temperament able to deal with the vagaries of chance. Evidence suggests that at least 95 per cent of us just don't have what it takes.

Insure it – or chance it?

Whether we like it or not, we all have to make bets. They may not involve a casino or a bookmaker, but they still involve money and uncertainty. If you've got a house, you'll have buildings insurance, and probably cover for the contents as well. In other words, you're handing over a sizeable sum reflecting your view of an uncertain event: some calamity striking your home. That's a bet – as is health insurance, life assurance and making investments. But are they good bets? That's a question that probably crosses the minds of most people who've bought consumer electronics products and been offered an 'extended warranty'. Once only available with big-ticket items, by the mid-1990s they were being offered on just about everything, from phones to fryers. They're still big business today: in the UK alone, millions of people take up extended warranty offers each year, and hand over around £1 billion in premiums. But there's also been a lot of controversy over whether they're actually worth the money. Some insist that the failure rate of most electronic products is far too low to justify the premiums being charged. Others argue it's all a bit more complicated than a question of probabilities: those who take up extended warranties are buying peace of mind as well as cover for replacing the product.

The truth is that the bets we make in real life are indeed more subtle than those made in casinos and the like. Fortunately, the basic concepts for understanding them were developed centuries ago. The result is one of the most useful applications of the laws of chance – and also one of the most controversial. The starting point for every decision made in the face of uncertainty is a question: what are the likely consequences? The basic method for answering this was developed by the brilliant seventeenth-century French polymath and pioneer of probability theory Blaise Pascal. For something so powerful, it's strikingly simple: the consequences we should expect from an uncertain event can be gauged by multiplying those consequences by the chances of the event actually happening. Suppose, for example, we're offered a bet with a 20 per cent chance of winning £100. The £100 is the consequence of the bet going our way, so according to Pascal's argument, the consequences we should expect to get from this uncertain event are £100 times the 20 per cent chance of it happening, giving an expected value of £20. All very simple – but does this make sense? After all, we'd never actually win a payout of £20; we'd get either £100 or flat nothing. True, you know this only after you've made the bet, and it's a bit late by then. The beauty of Pascal's rule is that it allows us to gauge the worth of the gamble *before actually making it*. To see this, imagine that over the course of your life you faced a very large number of these '1 chance in 5'-type bets – so large that the Law of Large Numbers is pretty reliable. Then we know that we'd win pretty close to 20 per cent of all of them, and so on average we'd take home 20 per cent of all those £100 prizes we were offered. Pascal's rule simply applies the same reasoning down to each individual bet. And in doing so, it gives us a number – the *expected value* – that allows us to decide whether any bet is worth taking, ahead of time. We just have to ask ourselves if the expected value of winning is worth the cost of taking part.

In the case of our 20 per cent bet, we've worked out that the expected value of winning is 20 per cent of £100, or £20. But we mustn't fall into the trap of so many amateur gamblers and be mesmerised by the prospect of winning; we must also face the possibility that we may lose – and there's an 80 per cent risk of that happening. So let's now apply Pascal's rule again, this time to our losses. Clearly, we don't want our expected losses to exceed our expected winnings, because that means we'll lose money in the long run. In our example, we can do this by ensuring we don't risk so much money that losing 80 per cent of it exceeds our expected winnings, which we already know is £20. So that means we mustn't risk more than £25 (as 80 per cent of that equals £20). Sure, you may get away with doing so once, or even a few times, but keep breaking Pascal's rule, and you'll end up sorry.

The power of Pascal's rule can be applied to more than silly games. For professional gamblers, it is the light that guides them towards serious money, and underpins the Golden Rule of Gambling. In that case, we're trying to tell whether the payoff (in the form of the offered odds) is reasonable, given our judgement about the chances of winning. If the expected value of a win exceeds the expected cost of a loss by a comfortable margin, then we've got ourselves a 'value bet'. Expected values are also crucial for assessing the 'bets' we're faced with making in that great cosmic casino we call Life. Take the case of extended warranties. In 2013, the UK Consumers Association magazine *Which?* examined what it called 'The great extended warranties rip-off'. Its investigation focused on stores giving inaccurate information about the warranties, and concluded that the warranties weren't worth the money, which probably came as little surprise to many. Yet while the magazine made some hand-waving statements to back up its conclusion, it failed to show just how big a rip-off extended warranties are. Doing that is a nice exercise in the use of expected value. In the *Which?* survey, one supermarket was found to be charging £99 for five-year coverage for a TV worth £349 new. Now, if you've just parted with that amount of money, £99 may not seem like much more to pay to get five more years of peace of mind. Yet the application of expected value theory may give you pause. If the TV fails, the 'expected loss' is the cost of the TV multiplied by the chances of the failure – which we don't know. What we do know, however, is that this expected value shouldn't be bigger than the £99 insurance premium we're being asked to pay – because then we're paying over the odds for the risk. So that tells us that the warranty is worthwhile only if it's less than £349 multiplied by the risk of failure. That in turn means the risk of the TV breaking down over five years needs to be *at least*

99/349 = 28 per cent. If you think that's reasonable, then go ahead, but you might want to check out the true failure rate – as *Which?* did. The actual failure rate is just 5 per cent. That's way below the minimum failure rate for the £99 premium to be fair. It also allows us to calculate what a fair premium should be: it should be £349 times that 5 per cent failure rate, or about £18 – just a fraction of the £99 being charged.

The TV warranty wasn't even the worst offender: one chain of electronic stores offered a £139 'premier' five-year warranty for a £269 TV – which sounds ludicrous even without doing the maths. But given that the true failure rate was just 2 per cent, it means a fair premium would be a staggering 26 times less than that actually charged. With profit margins like those, it's no surprise *Which?* found so many retailers keen to push extended warranties on us – or at least those of us who can't do the maths. Now we can, thanks to Pascal's rule of expected value. That rule shows that we're being charged over the odds for insurance if the premium greatly exceeds the value of the product multiplied by the chance of it going wrong over the period insured. In the case of TVs at least, the failure rate is just a few per cent, and so a fair premium should not be more than a few per cent of the purchase price (and even that ignores depreciation – which with technology is horrendous).

The same basic idea can be used when getting insurance against losing a gadget, or having it stolen: a reasonable premium to pay is roughly the value of the gadget, times the chances of the event occurring. Crime statistics are well worth checking out here, as they often reveal that a premium of more than a few per cent of the value of the gadget is a complete rip-off. In the absence of hard statistics, personal experience can help gauge the risks. The very fact something has *not* happened to you is surprisingly informative. A bit of maths shows that if an event has never occurred despite being given N opportunities to do so, then one can be pretty confident the frequency of the event occurring is no more than 3 divided by N. So, for example, if over the last five years you've never lost any possessions in circumstances similar to those in which you plan to use your new gadget, the chances of it being the first are likely to be less than around 3/N, where N is the number of relevant possessions. If you reckon to have more than a few dozen of these, then 3/N is around 10 per cent, and a fair premium to pay over five years would thus be no more than 10 per cent of the price, giving an annual premium of around 2 per cent of the purchase price.

Some people (especially those working in insurance) will protest that all this is hopelessly simplistic. And in some respects, it certainly is. Most obviously, we've ignored the fact that insurance often provides more than just the cost of replacement; many policies include services like 24-hour on-site repair. Peace of mind and convenience are worth paying for, even if they're hard to quantify. Then there's the problem of being able to cope with the consequences if your 'bet' about the need for insurance goes wrong – which, as we're dealing with uncertain events, is always possible. It makes perfect sense to decline insurance demanding ten times what you regard to be a fair premium *if* you can cope with the consequences should your entirely rational decision prove wrong. If it's a gadget, say, or a washing machine, then the cost may be annoying but not catastrophic. It's a different matter with insurance for your house, say, or for medical cover for trips abroad. You might well think the risk of you falling ill on a short trip is so low that paying (say) a £20 premium isn't worth it. But with hospital bills and repatriation costs quite capable of exceeding 10,000 times that amount, are you *really* confident betting that the chances of some mishap really are less than 1 in 10,000, when the cost of losing the bet is an eye-watering £200,000?

This highlights a key fact about insurance, and indeed decision-making in general: context is everything. If you're poor, even a fair premium may be beyond your means; regardless of how rational you are, you have no choice but to trust to luck. On the other hand, wealthy people may actually be willing to pay more than the fair premium, simply because money means less to them. The fact is that while the percentage may be the same, a multimillionaire paying £10 million out of £100 million won't feel as much pain as someone on benefits handing over £10 from £100.

This dependency of the value of money on context is crucial to making decisions about it, and was spotted by the pioneers of probability theory. By the early eighteenth century, Pascal's rule for making decisions using expected value had become widely known. It seemed to imply that all decisions involving money could be made by multiplying the probability of each outcome by the amount of money involved. But in 1713, the Swiss mathematician Nicolaus Bernoulli (whose uncle was Jacob Bernoulli, of Golden Theorem fame) pointed out a problem. Put simply, he pointed out that Pascal's rule could lead people to make utterly unrealistic decisions. For example, imagine being asked to take part in a game where a coin is tossed and heads constitutes a win – and, to make thing a bit more interesting, the jackpot doubles on every toss until heads finally comes up. How much should you be willing to pay to take part? Pascal's rule says it's the 'expected value' of playing, which is the probability of winning a toss – that is, 50 per cent – multiplied by the amount of money on offer, which doubles with each toss. Now clearly the longer the game goes on, the bigger the win, but also the bigger the chances of the game stopping. Applying Pascal's rule, these two counterbalancing effects lead to an expected value from playing of … infinity. So the decision is clear: you should sell everything you have to play the game, as the expected winnings are infinite. Yet as Bernoulli pointed out, this is patently ridiculous. First, the chances of recouping the infinite entry fee in winnings are essentially zero. In order to win even the modest sum of £16 we'd need the coin to come up tails four times before ending with heads, and there's only a 3 per cent chance of it doing that. Then there's the small matter that the organisers of the game will have only a finite amount of money to pay us anyway. Yet Pascal's rule tells us it still makes sense to ignore all this and pony up an infinite amount.

This bizarre result became known as the St Petersburg Paradox, as the person who resolved it (mathematician Daniel Bernoulli, cousin of Nicolas) revealed his solution to the St Petersburg Academy of Sciences in 1738. While the problem itself seems like one of those silly mind-games loved by academics, it prompted

Bernoulli to invent a concept that today underpins the $100 billion global insurance business: utility. In doing so, he formed a striking connection between the cold, platonic world of mathematics and the warm, fuzzy world of human psychology.

According to Bernoulli, the paradox exists only because Pascal's rule focused purely on arithmetic, and failed to consider the subjective notion of the *value* of money. This, Bernoulli argued, depends crucially on context – and specifically how much of the stuff we have. A billionaire sees less value – or 'utility', in the jargon – in £100,000 than someone on welfare. But even the billionaire can see some utility in it. Bernoulli therefore proposed that when making decisions involving money, Pascal's rule should give us not the expected monetary value of the consequences, but the expected *utility*. But what is that – and how do we calculate it for a given amount of money? Bernoulli derived a simple conversion rule, based on his argument that while extra money always adds *some* extra utility, the effect tails off as we get more of it. Mathematically, this implies that the utility of a given amount of money is proportional to its logarithm. So, for example, £1,000 has a utility of 3 units (known as *utiles*) because the logarithm of 1,000 is 3, while £1 million – a sum 1,000 times larger – has only 3 units of extra utility, as the logarithm of a million is 6. This, Bernoulli argued, has a big impact on how people with different levels of wealth will view monetary decisions. Seen through the prism of utility, if you have £1,000 and are offered a chance to win another £1,000, that equates to a jump in utility from 3 to 3.3 *utiles*, as the log of 2,000 is 3.3. In contrast, someone with £1 million already has a utility of 6, and winning an extra £1,000 means increasing their financial utility to the log of 1,001,000 – that is, a rise from 6 utiles to 6.0004, which hardly seems worth it.

While one can argue over the precise 'conversion rate' of money to utility, the key point is that they don't change pro rata: utility grows ever more slowly with increasing wealth. And it's that which can turn the 'silly' St Petersburg Paradox into something sensible. If we use Pascal's rule and swap expected winnings for expected *utility* of taking part, the result is dramatic. As the number of tosses increases, the expected utility no longer grows ever larger, but instead settles down to a sensible finite value – and clearly there's no point paying an infinite amount to win that.

Over the years, academics have argued over the merits of Bernoulli's resolution of the paradox, and its lessons for real-life decisions. After all, it's hard to imagine anyone being stupid enough to consider paying an infinite amount of money for anything (hard, but not impossible – see box opposite). What isn't in doubt is the transformative effect all this has on the hard-nosed business of insurance – as Bernoulli himself realised. Offering to compensate people for their misfortune is a lovely idea, but it does have to make financial sense. And given that we're dealing with uncertain events, that's not a trivial problem. For a start, insurers will want to ensure there's enough money coming in via premiums to cover the misfortunes. That means estimating the likely risk and then setting premiums a bit bigger than the likely financial hit, using the same logic that leads bookmakers to offer payouts lower than the odds suggest is fair. Insuring lots of people also helps, as the Law of Averages will then drive the frequency of misfortunes closer to the expected rate – in theory at least.

Bernoulli's concept of utility leads to many other, more subtle consequences, such as showing that insurers who share risks with rivals can both reduce their own exposure and offer lower premiums to their clients – so everyone wins. Put simply, Bernoulli's theory allows insurers to cover many risks they'd otherwise have to turn down. That's generally a good thing, allowing us to buy peace of mind for everything from cancelled holidays to broken boilers. But others may see it as a way of exploiting our neuroses. Certainly that's what those extended warranty policies appear to be. In reality, insurance companies know that Bernoulli's concept of utility has only limited, well, utility. Most obviously, if a risk is too high or vague, the gap between the fair premium and one clients will pay becomes too small to be commercially worthwhile. Public liability cover often falls into this category: the risks are hard to judge, and the payouts can be colossal. This has compelled the insurance industry to develop a variety of techniques to

How the St Petersburg Paradox cost the world $5 trillion

In 1957, David Durand, a finance professor at the Massachusetts Institute of Technology, pointed out the disturbing parallels between the 'absurd' game that Bernoulli tackled, and investment in so-called growth stocks. These are shares in companies whose revenues seem to be skyrocketing. Such companies routinely make headlines in the media, sparking huge interest from investors. While many just pile in anyway, serious investors prefer to probe further, to find out if the stock price can really be justified by the company's prospects. Put simply, that involves estimating the current worth of the company's future performance and assets, assuming certain growth rates and interest rates. The problem, of course, is that no one knows for certain what those future rates will be. Worse still, this so-called 'discounting' process assumes that the company goes on and on for ever – just like the game at the heart of the St Petersburg Paradox. Financial analysts who assume growth rates will never fall below interest rates then end up with valuations consistent with share prices of … infinity. Surely no one would be stupid enough to believe such 'analysis'. Think again. A study published in 2004 by mathematicians Gabor Székely and Donald Richards concluded that a St Petersburg Paradox phenomenon was a key driver of the notorious 'Dotcom Bubble' of the late 1990s. The 'growth stocks' were high-tech companies which had never turned a profit but whose stock soared to levels defying common sense – but entirely consistent with the crazy valuations. When it burst, the Dotcom Bubble wiped $5 trillion off the NASDAQ, the US stock market where the shares were traded. Still, we should count ourselves lucky; it could have been more – infinitely more, in fact.

allow it to offer cover against the vicissitudes of life. Some are very simple – such as only covering losses above a certain minimum, or 'excess'. Others are the product of insights into probability that allow insurers to take on truly extraordinary risks, such as Extreme Value Theory, which we'll encounter in a later chapter.

Like casinos, insurance companies have built their business model on the laws of probability – and it works very well for them. And most of the time, it works for us as well, though we may suspect we're sometimes being ripped off. Yet we don't have to make a stark choice between taking out insurance or just chancing it. There is a middle way – at least for small items, which just happen to be where many of the rip-offs lie. Pascal's rule allows us to be our own insurers. We just work out a fair premium by multiplying the value of the item by the risk of calamity, and then pay this in instalments into our own sinking fund. Alternatively, we can simply pay in the premiums that would otherwise have gone to the insurer – which you can be sure will be more than enough. Either way, we're then covered if calamity strikes, or end up with a nice nest egg if it doesn't.

Of course, it makes sense to be pessimistic. It's possible that several calamities will strike at once before all the premiums have been paid, so one should put a decent amount into the sinking fund to cover that possibility. We must also never lose sight of the purpose of the sinking fund: it's there to be raided if and only if calamity strikes. Sure, we'll feel really irritated if our brilliant ruse doesn't pay off, but that's something we just have to live with: as we'll see in the next chapter, making decisions about risks is not always rational. As they deal with probability, Pascal's rule and utility theory cannot give guarantees – and we must make provision for if the best-laid plan goes wrong. But as a means of keeping money in our pocket rather than giving it to The Man, they're priceless.

↑UPSHOT

We live in a risky world, and insurance was invented to help us deal with the consequences – while turning a profit for insurance companies. Simple rules of thumb show when it's worth taking out insurance, when we're better off chancing it – and how to make provision for when even the best-laid plan goes wrong.

Making better bets in the Casino of Life

I s it worth asking the boss for a raise? Should we act on the rumours about how the neighbourhood is going to change? What's the best way to deal with global warming? Every day we're faced with making decisions, or at least having a view on them. Yet even the minor ones often seem overwhelming, with their multiple uncertainties and consequences. Combine those with the fear of making a bad decision, and it's no surprise that we often simply decide not to decide. Fortunately, making decisions in the face of uncertainty has long been a big part of the theory of chance, resulting in a range of tools that cut through the complexity. They're remarkable for their power to extract insights about big questions with little effort. The originator of what's now called decision theory, the brilliant French polymath Blaise Pascal, used it to tackle one of the definitive Big Questions: does it make sense to believe in God?

Contrary to what is sometimes claimed, Pascal was not attempting a proof of the existence of God. In his view, God is so ineffable and incomprehensible that any such proof – or indeed disproof – was unlikely to mean much. A question that was worth asking, Pascal argued, was whether it made sense to believe in God. He began by returning to his concept of expectation, according to which it's not just probabilities of outcomes that matter, but their corresponding consequences. As for what those consequences are, Pascal was pretty vague, but the essence of them can be summed up in the following table.

	God exists	God does not exist
Choose to believe	*Consequence*: positive – eternity in paradise	*Consequence*: negative – wasted time and effort on rituals
Choose not to believe	*Consequence*: negative – potentially big trouble from wrathful God	*Consequence*: positive – saved some time and effort on rituals

Why belief in God makes sense, according to Blaise Pascal

Note that, in contrast to a simple bet, we're no longer faced with a straightforward 'win/lose' outcome. Instead of having to decide whether God does or does not exist, Pascal has shown it's possible to cope with more complex situations involving both possibilities. As the table shows, there are now four scenarios to deal with. To decide which choice is best, Pascal suggested we work out the *expected* consequences of each. That means multiplying each consequence by the appropriate probability. But how are we supposed to estimate the chances of the existence of God? Apparently arguing that there was no way that reason could prefer one over the other, Pascal opted to set them as equal: 50:50. You don't have to be an atheist to think there's something wrong about this; after all, if you knew nothing about a horse, would you blithely assume it had an evens chance of winning? Pascal was wrestling with a difficulty that still causes problems today: what probability to give to something about which you know nothing. We'll encounter this again in other contexts, but for now let's just move on – as Pascal is about to use a trick that short-circuits the whole problem anyway. Assuming for the moment there really is an equal chance of God existing or not existing, then the probabilities are the same in all cases, and their impact cancels out. We're left with simply comparing the consequences of each choice, and seeing which comes out best. According to the right-hand column in Pascal's table, the best outcome on offer if God does not exist is merely some saved time and effort. On the other hand, the best outcome on offer if God does exist is eternity in paradise. So by that reckoning, belief in God makes perfect sense. Or at least it does according to the way Pascal has set it up; but what if we don't buy his argument of a 50:50 chance of God existing? Now we have to work out the four *expected* consequences in full, multiplying each individual consequence by its associated probability and seeing which combination comes out on top. All very tedious and problematic. But as a mathematician Pascal knew there was a way to avoid all this. He simply declared that the consequences of believing in a God who does exist – namely, *eternal* life in paradise – are not merely positive, but actually *infinite*. With all the other consequences being merely finite, it doesn't matter what the various probabilities are: the decision leading to the only infinite payoff always wins. At a stroke, Pascal thus made belief in God the only rational decision.

Again, if you think this is all deeply suspicious, you're in good company; few scholars take Pascal's argument seriously today, for all the reasons given and more. What we should all take seriously, however, is Pascal's basic approach to deciding between various options. Used with less artifice, it can cut through a lot of complexity to give us clear-cut decisions in the face of uncertainty. We don't even have to go so far as doing any sums; simply writing out a table like that for Pascal's wager can often help clarify the best course of action.

Suppose a manufacturing company is trying to decide how to react to news that a chemical it has been using may be bad for the environment. The problem is that the evidence isn't very convincing, and may not stand the test of time. So the company is facing making a decision under uncertainty. So let's create a table of the various consequences (see opposite).

It's not going to be easy to turn these consequences into numbers and then multiply them by the unknown probability the chemical *really is* toxic. So let's take a leaf out of Pascal's book and see if it can be avoided. We don't have to go as far as he did, and bring infinity into the mix. Instead, we can simply look for what's called 'dominance' – that is, seeing if one decision is better

	Chemical is toxic	Chemical is not toxic
Decision A: Keep using chemical	*Consequence*: environmentally harmful; lawsuits, bad publicity.	*Consequence*: Business as usual, but may look complacent.
Decision B: Switch to a substitute	*Consequence*: good for the environment, good for company's image.	*Consequence*: unnecessary changes, but would look responsible.

How should the company respond to the claims about its product?

regardless of the probabilities. In the example, it's obvious that *if* the chemical really does prove toxic, then switching to a substitute is the better decision. Rather trickier is choosing between the consequences of each decision if the chemical does *not* prove toxic. But if we reckon the upheaval and cost of changing aren't too big and could easily be justified by the resulting image benefits, then it's clear that switching is still the best thing to do. As it gives the best consequences *regardless* of whether the chemical is toxic or not, we no longer have to worry about pinning down the exact probability: switching to the substitute is always better.

Every case has to be taken on its merits, of course, but the fact remains that sometimes there's a dominant strategy that make it possible to reach the best decision without fretting about the chances involved. Usually, however, we do have to plug in some numbers to capture the relative merits of the various consequences. It doesn't matter what the range is: −10 to +10 for the worst to the best outcome is as good as any. So, for example, a family may be contemplating moving house after hearing of rumours that a new road is to be built near by, and after discussion has come up with the following analysis and relative benefit scores for the various consequences:

	Rumours are true	Rumours are false
Decision A: Stay put	*Consequence*: noisy and unsafe location, plus hard to sell house. **SCORE: −10**	*Consequence*: business as usual. **SCORE: +7**
Decision B: Move house	*Consequence*: no road menace, but longer commute and trips to school. **SCORE: +2**	*Consequence*: unnecessary upheaval and expense, but maybe time to move. **SCORE: +1**

Unlike the company dealing with the chemical scare, the family can't opt for a sure-fire decision that's best regardless of whether the rumours are true or not. To make their decision, they need to compare the *expected* consequences of each decision, and that requires some estimate of the probability that the rumours are true. But once again, we can dodge that tricky problem. This time, while the probabilities are needed to make a decision, we don't have to specify them. Instead, we can flip the problem around and ask what does the probability *need to be* before it makes sense to move? Some simple maths[1] shows that, in this example, moving house makes sense if the family believes there's a greater than 1 in 3 chance of the rumours being true. If that sounds implausible, they should stay put.

If making so big a decision based on numbers like these makes you feel uncomfortable, then consider the alternative: going with gut instinct. That exposes us to the danger of making our decisions on the basis of factors which have emotional impact, but which are actually irrelevant. If you think you're immune to such human weaknesses, imagine yourself as the hard-nosed CEO of a company with 450 staff going through tough times. You know you're probably going to have to scale the business down, so you're facing decisions that could have a big impact on your workforce. Keen to do the best by them, you've hired a top firm of management consultants to decide on the way forward. In time-honoured fashion, they have handed over a hefty report, an invoice to match – but no recommendation. Instead, they've come up with two options.

Plan A1: Restructuring the company, saving 150 jobs;
Plan A2: Do nothing, which carries a 2 in 3 chance of complete closure and a 1 in 3 chance that 450 jobs will be saved.

So, which do you go for? If you're like most people, you went for Plan A1, with the relative certainty of the 150 jobs. But, mindful of the huge impact your decision will have on your workforce, you get a second opinion from a second management consultancy company to ensure you've covered all the options. The result is another hefty report and invoice – and still no recommendation. But again they offer two plans:

Plan B1: Continuing as normal, leading to the loss of 300 jobs;
Plan B2: Restructuring the company, giving a 1 in 3 chance that everyone will keep their jobs, but a 2 in 3 chance that 450 jobs will go.

Now which one looks better? Plan B1 looks really awful, while Plan B2 does seem to hold out some hope. So now it's just a matter of deciding whether to go with Plan A1 or Plan B2 – and retaining the appropriate management consultants to bring about the restructuring. Or is it? While doing the comparison, you may start to notice something odd. Isn't Plan A1's promise to save 150 jobs out of the total workforce of 450 via restructuring the same as Plan B1's dire warning that continuing as normal will lead to 300 redundancies? Applying a bit of Pascal's theory reveals another odd fact: Plan B2's promise of a 1 in 3 chance that all 450 staff will keep their jobs implies an expected loss of $450 \times 1/3 = 150$. That's just what Plan A2 offers. In short, in terms of likely redundancies, all four plans are identical. All that's different is the way they've been presented. Plan A1 emphasised the certainty of a good outcome, whereas Plan B1 linked certainty to the bad outcome. And as Nobel Prize-winning research by Daniel Kahneman and Amos Tversky has shown, people faced with decisions prefer certainties over gambles whenever an outcome is good – that is, they become risk averse, preferring a guaranteed upside. Yet if the outcome on offer seems bad, people will suddenly become risk-seeking and happily take a punt on getting the positive outcome. Anyone aware of these human traits can nudge others towards a particular decision simply by presenting it in the right way. An unscrupulous consultant (imagine!) wanting a client to choose a particular plan should emphasise any certainties it gives of positive outcomes – while focusing on the guaranteed downsides but uncertain upsides of the alternatives.

Using decision theory helps inoculate us against such trickery, forcing us to do the cold, hard maths. As we've seen, sometimes there's no maths to do: one set of consequences dominates the other, regardless of what actually happens. We saw how such dominance helped the company dealing with the potential threat from the use of an allegedly risky chemical. But it can also be applied to much bigger issues as well. For instance, one of the biggest current challenges facing the world is how to deal with the threat from global warming. Some argue for drastic measures, such as the complete abandonment of fossil fuels. Others think we should focus on adapting to the changing climate, while still others insist global warming is a myth – or at least, nothing to do with human actions. There are some very good reasons for believing that global warming is under way, and with it climate change across the planet. So what should we do? Decision theory again helps cut through the complexities to lay out the options in an incontrovertible way. After all, even the most diehard environmentalist or climate change sceptic can at least agree that global warming is either a reality or a myth. Setting up the usual decision matrix shows that all governments should focus on cutting energy consumption through improved energy efficiency, as it's a dominant strategy – that is, it makes sense, regardless of the realities of global warming (see table opposite).

It's a conclusion that's now being endorsed by the likes of the International Energy Agency and the United Nations Foundation, which describes energy conservation as 'the first and best step toward fighting global warming'. Yet for decades it has remained the Cinderella of global energy strategy, ignored by politicians. Perhaps someone should give them a primer in decision theory.

	Global warming is real	Global warming is a myth
Decision A: Cut energy consumption through efficiency drive	*Consequence*: Upfront costs, but delay/avert substantial impacts; more resources/money for amelioration; better energy security.	*Consequence*: Upfront costs, but conserves resources and money, and improves energy security.
Decision B: Do nothing	*Consequence*: No upfront costs, but major impact on health, economy, global security, etc.	*Consequence*: No upfront costs, but no future savings in resources or money, or better energy security.

Why energy conservation is a no-brainer way to combat global warming

↑UPSHOT

Decisions often involve a tricky mix of unclear probabilities and big consequences. Simply writing down the mix of possibilities and consequences in a table can make the best course of action clear. If not, the basic arithmetic of decision theory is always worth trying.

Tell me straight, doc – what are my chances?

When Alice started feeling pain in her left breast, she took no chances. As a woman in her sixties, she was already undergoing biennial mammograms – and she decided to bring her next one forward, to find out the truth as soon as possible.[1] The X-rays done, she felt good about doing the right thing as she left the medical centre, with the receptionist saying they'd call if there was a problem. But a few days later, the medical centre did call – and not to give Alice the all-clear. It seemed the mammogram had come up positive. Alice was deeply concerned; who wouldn't be? A quick trawl of the internet reveals that mammography is around 80 per cent accurate. The implication seems clear: there's an 80 per cent chance that Alice has got breast cancer. That's certainly what many doctors would conclude.[2] But they'd be wrong – along with, most likely, that positive mammogram result. That's not because the 80 per cent figure is wrong. It's because it tells only part of the story – and one inadequately summarised by the simple-seeming notion of 'accuracy'.

To make sense of any diagnosis, probability theory reveals we need not one but *three* numbers. Two of these reflect a key feature of any diagnostic test: its potential to mislead in two distinct ways. First, it can wrongly detect something that does not, in fact, exist – producing a so-called false positive. But the test may also miss something that really does exist, leading to a false negative. The ability of a test to avoid these two failings is summed up by two numbers: the true positive rate and the true negative rate – known technically (and with typical opaqueness) as the *sensitivity* and the *specificity*. Over the years, attempts have been made to combine these into a single number with some claim to represent 'accuracy', but they all fall short in some way or other. Keeping them separate, on the other hand, allows us to gauge how impressed we should be by a diagnosis. After all, any doctor can diagnose, say, heart disease in a way that's guaranteed to catch every case: simply tell every patient they have heart disease. The true positive rate will be an impressive 100 per cent. Yet obviously that's not exactly useful for diagnosis – and that's reflected in the fact that the true negative rate (specificity) is zero, because the doctor never tells anyone they *don't* have heart disease. The real value of a diagnostic test can only be gauged by knowing *both* rates individually.

In the case of mammography, they're both around 80 per cent. That means that out of 100 women with breast cancer, mammography would correctly diagnose the disease in around 80 of them, while out of 100 women free of the disease, around 80 would correctly be given the all-clear. That may still seem unnervingly reliable, but as so often in matters of probability, the precise wording is critical. The 80 per cent reliability figure comes from tests on women whose breast cancer status was already known. As such, it measures only the reliability of the test in confirming what's already known. But for women like Alice undergoing routine screening, all we know about her breast cancer status beforehand comes from estimates of the *prevalence* of the disease (or 'base rate'). This is the crucial third figure we need to make sense of the outcome of any diagnostic test – and its impact can be dramatic.

Again, take the case of Alice. The prevalence of breast cancer depends on a host of factors, from ethnic background and genetic profile to age, and to make sense of any individual test result it's vital to use the appropriate figure. For example, the lifetime risk for women in the USA is around 12 per cent, but that's skewed by the huge increase in risk with age. For a 60-something woman like Alice, the prevalence is around 5 per cent – a figure that radically alters the implications of a positive result from the '80 per cent

What an 'accurate' test result *really* means

As a diagnostic technique for breast cancer, mammography is pretty impressive: it will detect around 80 per cent of breast cancer cases, and will give the all-clear to a similar proportion of those who are free of the disease. But that tells us precisely nothing about the probability that Alice has breast cancer, given her positive test result – because we don't know which of these two camps she falls into. We can, however, get some idea of this from the prevalence of breast cancer among women like her. Statistics show that the risk figure for women in her age-group is around 1 in 20. So let's look at the implications in raw figures. Out of 100 women like Alice:

Number with breast cancer: 5
Number without: 95

Of the five with breast cancer, the true positive ('sensitivity') of the test will detect around 80 per cent, or four women. But crucially, they aren't the only ones who will get a positive result. Of those who are free of the disease, the true negative rate ('specificity') of 80 per cent means that most will correctly be given the all-clear – but there will still be 20 per cent who won't. Combined with the fact that 95 per cent of the women don't have breast cancer, that leads to an awful lot of false positives:

Number of correct positive results: 80% of 5 = 4
Number of incorrect positive results: 20% of 95 = 19
So total number of positive results: 4 + 19 = 23

So now we can finally answer the key question Alice had about her positive test result: what are the chances she really does have cancer?

Pr (Breast cancer, given positive result) = no. of *true* positives/ total no. of *all* positives = 4/23 = 17 per cent!

So it's over 100 – 17 = 83 per cent likely that Alice is *free* of breast cancer, despite the positive mammogram.

accurate' mammogram. Some simple arithmetic (see box above) reveals that in fact it's over 80 per cent likely that the positive result is actually a false alarm.

That's pretty much the exact opposite of the apparent implications of getting a positive result from a test described as '80 per cent accurate' – and shows the critical importance of taking into account the plausibility of any diagnostic test result.

So how should one react to a positive result? Certainly, it makes sense to be somewhat concerned: in Alice's case, for instance, the positive test result increased the chances of her having breast cancer from 5 per cent – the 'base rate' for her age-group – to 17 per cent. But there's certainly no cause for fatalism or panic, as even that higher figure still means there's an 83 per cent probability that it's not breast cancer. The appropriate response is to undergo further tests, as each one adds evidential weight for or against a diagnosis of breast cancer. This is exactly what Alice did – and sure enough, she got the all-clear.

It doesn't always work out this way, however. Probabilities aren't certainties, and one should never push them too far. When she detected a lump in her breast, singer Olivia Newton-John was still in her early forties – and thus had a breast cancer risk of barely 1 per cent. A mammogram test came up negative, as did a biopsy. Even so, she felt increasingly unwell and it turned out she did have breast cancer after all. Fewer than one in 10,000 women her age would be so unlucky as to have two false negative results. Yet probability theory tells us that, given enough opportunities, even low-probability events manifest themselves. It's just that we rarely hear of them.

The same reasoning also shows that every woman undergoing regular screening should prepare for at least one scare. The flip-side of 80 per cent reliability at ruling out those who really don't have cancer is a 20 per cent risk of false positives. Over the course of a dozen or so biennial tests over the age of 50, that implies a high chance of experiencing at least one scare.

With ever more diagnostic tests emerging from research labs, the need to know how to interpret them has never been more important. Yet all too often even the researchers prefer to highlight some more or less meaningless measure of 'accuracy', while the role of base rates is ignored altogether. In July 2014, researchers at two leading UK universities unveiled a blood test said to be '87 per cent accurate' in predicting the onset of Alzheimer's disease among people with mild memory problems. The story made headlines in the media, and was hailed as a major breakthrough by UK govern-ment health secretary Jeremy Hunt. Some researchers felt the need to put the story in context, however, and one expert on Alzheimer's cautioned that the impressive 'accuracy' figure still meant around one in ten patients would get an incorrect diagnosis.

In reality, it's unclear what the figure meant, as the research-ers themselves never made explicit what they meant by 'accuracy'. That said, they took great pains to establish the two failure modes of the test, as reflected in its sensitivity and specificity. Using data gathered from hundreds of patients with various forms of demen-tia, they found that the blood test correctly predicted progression to full-blown Alzheimer's in around 85 per cent of cases, while cor-rectly predicting no progression around 88 per cent of the time. These figures imply a false negative rate of 15 per cent and a false positive rate of 12 per cent. But as with mammograms, we can only make sense of a positive test result if we can assess its plausibility – which means knowing the base-rate risk of Alzheimer's among those who take the test. As it was devised for use with people who already had mild cognitive impairment, the relevant base-rate risk is around 10–15 per cent. Cranking through the same arithme-tic we used above to interpret mammograms, it turns out that a positive Alzheimer's blood test implies barely a 50:50 chance of progressing to Alzheimer's disease. So, as with mammography, the 'accuracy' figure looks less impressive when put in context. Cynics might claim the test is no better than a coin-toss, but that's unfair. In raising the probability of progression from 10–15 per cent to 50 per cent, the blood test has undoubtedly added genuine evidential weight, which a coin-toss can never do. As such, it may one day become part of a battery of tests for Alzheimer's, just as mammography and biopsy have for breast cancer. But the fact remains that there's a big difference between the 'accuracy' of the test and the odds of Alzheimer's implied by a positive test result.

The dangers of misinterpretation are most acute among those who decide to test themselves using home diagnostic kits. First introduced in the 1970s for pregnancy testing, it's now possible to buy home testing kits for many conditions, from allergies to infection with the AIDS virus, HIV. As ever, they're said to be impressively 'accurate', but exactly what that means and in what context is often far from clear. In the case of home pregnancy tests, the claims of accuracy can be taken pretty much at face value: if it's positive, it's highly likely you're pregnant. Such tests have extremely low false positive and false negative rates – and, in addition, most women taking such tests already have strong reasons for believing they're pregnant. Yet even a pregnancy test can prove pretty unreliable if taken by someone unlikely to be pregnant – such as a man. In 2012, one user of the social media site Reddit reported how a male friend had used a spare pregnancy test left by his girlfriend as a joke – and was stunned to get a positive result.[3] As the base rate of pregnancy among men is rather low, he wasn't likely to give birth, despite the outcome from the 'accurate' test. But that wasn't the end of the story. Other Reddit users pointed out that the test works by detecting the hormone HCG, which is produced by pregnancy in women – and testicular tumours in men. A visit to the doctor confirmed the diagnosis, leading to early treatment that may just have saved the 'pregnant' man's life.

The need to factor in the plausibility of diagnosis is especially important with home HIV diagnostic kits. These too are said to be well over 90 per cent 'accurate'. Yet unless you have very good reasons for believing you may have acquired the AIDS virus, that figure is dangerously misleading. While the specificity and sensitivity are indeed higher than 90 per cent, the base rate of HIV outside well-known risk groups is very low. As a result, positive test results for those outside known risk groups are far more likely to be false alarms than true positives.

It's not just medical diagnoses where the concept of accuracy needs to be considered with care. The same applies to any test claiming to exploit telltale signs of some trait – such as being a liar. Centuries ago in Asia, it was believed that dishonesty could be 'diagnosed' by stuffing the mouths of suspects with rice before interrogation. Those having the most trouble spitting it out after questioning were deemed guilty on the grounds that their deception gave them dry mouths. This sounds less than reliable, and what we've seen with medical diagnosis above crystallises those doubts: while the method may have a reasonable true positive rate, its false positive rate is also likely to be high, given that honest people are likely to have dry mouths through fear of not being believed. Then there's the need to put any positive result in context – which demands an estimate of the chances that the person is a liar, before applying the test.

Not that any of this has deterred people from claiming to have developed 'accurate' lie detectors. Since the 1920s, most attention has focused on so-called polygraphs, which monitor a host of physiological signs from heart rate to sweatiness believed to reveal when people are lying. Yet they've struggled to overcome the problem of false positives caused by stress, while being routinely fooled by trained spies. Aldrich Ames, the CIA analyst who worked for the Soviets during the 1980s and 1990s, passed repeated lie detector tests by following KGB techniques based simply on getting a good night's rest, staying calm and being nice to the polygraph examiners. That reduced the true positive rate, while false leads created by the KGB watered down the plausibility that Ames was a spy.

The quest to create a reliable lie detector continues unabated, however. In 2015, a joint team of British and Dutch researchers announced a technique based on the idea that guilty people fidget more. As ever, media reports focused on an impressive 'success rate' of 82 per cent, while being vague about what that meant. A draft paper by the researchers suggested it was actually just the average of the true positive rate of 89 per cent and a true negative rate of 75 per cent. If true, this is indeed a significant advance on the conventional polygraph, which according to the researchers has a typical strike-rate of around 60 per cent. Yet as ever these figures leave unanswered the key question: what are the chances of someone actually being a liar, given a positive test result? As we now know, that cannot be answered by the test results alone: we also need some insight into the chances that the suspect really is a liar, based on evidence from elsewhere. What we can say is that, if the 'fidget test' really is as reliable as claimed, those giving a positive result are still more likely than not to be honest unless there are reasons for thinking the chances of their being guilty exceed around 1 in 5.

Exactly the same problem faces everything from security screening at airports and fraud detection software to burglar alarms. While advocates focus on the supposed 'accuracy' of the technology, such claims are meaningless without some insight into the prevalence of what is being searched for. And if that prevalence is low – as, mercifully, it often is – then only incredibly high true positive and true negative rates can prevent a flood of false alarms.

There's a simple way of assessing the outcome of tests for rare events, which we'll call the Few Per Cent Rule.[4]

The 'Few Per Cent' Rule

If you test positive for something affecting less than a few per cent of people undergoing the test, then it's most probably a false alarm, unless the test is so good that its false positive rate is also below a few per cent.

It's clear we're all going to face false positives during our lives, from routine medical screening to bag searches at airports. That's not a reason for becoming blasé. Good decisions are based on combining probabilities with consequences, so a small chance of a devastating consequence should always be taken seriously. But nor is it reason for overreaction. In the end, our best protection against so much of what we fear is its sheer improbability. As the American industrialist Andrew Carnegie once wrote: 'I have been surrounded by troubles all my life long, but there is a curious thing about them: nine-tenths of them never happened.'

↑UPSHOT

When confronted with the outcome of a diagnostic test, don't be deceived by talk of its 'accuracy'. In many cases, the impressive-sounding figure is only half the story; many positive tests are more likely to prove wrong than right simply because of the rarity of what they're trying to detect.

This is not a drill! Repeat:
this is not a drill!

Living in one of the most earthquake-prone regions of the Earth, the inhabitants of Mexico City are understandably keen to have early warning of the next Big One. One day in July 2014, thousands of them got the news they'd been dreading. They had downloaded a mobile app which supposedly took data from Mexico's official seismic warning network. At around lunchtime, the app sent out a warning that a major earthquake was about to strike. Within seconds, people were rushing from their offices and onto the streets, bracing themselves for catastrophe. They waited, and waited – but nothing happened. Clearly, it had been a false alarm. The makers of the app issued an apology, saying they had misinterpreted a message from the official network. Then, barely eighteen hours later, the city was rocked by a strong earthquake of magnitude 6.3. The mobile phone app stayed silent.

Of all the natural calamities that scientists have sought to predict, none has defied them more resolutely than earthquakes. To this day, no reliable means of predicting the time, place and strength of an earthquake has ever been found. And it is not for lack of effort. The search for telltale signs ('precursors') of impending earthquakes dates back millennia. The Roman author Claudius Aelianus recounts how the inhabitants of the ancient Greek city of Helike noticed that mice, snakes and many other creatures left en masse five days before a catastrophic earthquake destroyed the city in the winter of 373 BC. Since then, claims for a host of other precursors have been made, from changes in groundwater and seepages of radioactive gas to changes in magnetic fields. And some of them have even been taken seriously. In the winter of 1975 the city of Haicheng in north-east China became the scene of strange events, with groundwater levels changing and snakes emerging from their lairs. Then the area was struck by a swarm of small tremors. Believing these to be foreshocks presaging a far bigger quake, Chinese geophysicists issued a warning that a major quake was about to strike, and a mass evacuation was ordered. On 4 February the earthquake struck: a devastating 7.3 Richter event. Yet out of the estimated one million people in harm's way, the evacuation led to all but 2,000 surviving. It seemed the dream of earthquake prediction had become a reality – until the following year when a 7.6 Richter earthquake struck the city of Tangshan. This time, there were no telltale foreshocks, and at least 255,000 perished. It was later claimed there had been strange animal behaviour in the area. Was this the vital clue that had been missed – or was it just post hoc rationalisation? How often do animals behave 'strangely' when everything is perfectly normal?

The answer seemed obvious: more research. The mission statement also seemed clear: if at first you don't succeed, try, try again. But that presumes that success is possible. What if it isn't? At the time of the Tangshan disaster, that wasn't a popular point of view. Many scientists kept faith with the dream that one day it would be possible to set up a network that detected precursors hours or even weeks ahead, allowing people to at least take shelter if not flee the area completely. Obviously, such precursors had to be reliable, and efforts to find them were redoubled. But one question attracted far less interest: just what level of reliability would be needed – and was any precursor ever likely to meet it?

Addressing this question meant seeing earthquake prediction for what it really is: a question of reliable diagnosis. And, just as medical diagnostic tests stand or fall by their ability to add weight of evidence about a specific risk, so too must any earthquake prediction method. If it is to be fit for purpose, it must be able to tell the difference between false alarms and the real thing. But crucially it must be able to do this well enough to be able to compensate for the fact that – mercifully – earthquakes big enough to merit mass evacuations are rare. The question then becomes: is that remotely plausible?

The outcome of applying the 'Few Per Cent Rule' of the previous chapter is not encouraging. It implies that if the risk of a big earthquake striking over the space of, say, a month is lower than a few per cent – which it certainly is – then an earthquake warning is most likely a false alarm unless its false positive rate is below a few per cent. That, in turn, demands that the precursors used by the system have such a low false positive rate. Despite centuries of effort, no precursor has ever been found that comes even remotely close to passing that standard. A more detailed analysis confirms this dismal truth.[1] Even if, by some miracle, a precursor was found with a 100 per cent true positive rate, the false positive rate would still need to be less than around 1 in 1,000 to compensate for the low probability of a major quake striking. Nothing that has been learned about the process by which earthquakes are triggered gives any hope of finding such a reliable precursor, either.

One might ask why such a basic problem with the very idea of earthquake forecasting did not kill off the whole quest decades ago. Cynics have pointed to the huge funds on offer to those willing to look for the elusive precursors. A more charitable explanation is that researchers were simply unaware of the probabilistic barrier that puts success forever beyond reach. The matter is now moot, as by the mid-1990s reality began to kick in. The abject failure of attempts to identify reliable precursors had become impossible to ignore. While some persist with the dream of predicting earthquakes as meteorologists predict storms, most seismologists have since joined one of two camps. The first accepts it's never going to be possible to predict earthquakes with the necessary precision in location and timing to permit action for any specific event. Instead, it focuses effort on an incontrovertible fact: that some parts of the world are at unacceptable risk of being hit by major earthquakes. There is no doubt, for example, that most of the most destructive ever recorded have occurred around the so-called Ring of Fire, which encircles the Pacific Ocean. It is also known for sure that some of these high-risk areas overlap areas with very high population density, notably Japan. Thus while no one can say exactly when or where the Big One will strike, we do know the locations facing a high risk of major casualties. Those seismologists in the first camp have made these high-reliability statements the basis for so-called *mitigation* strategies – such as making buildings and utilities more quake-resistant, and educating the public on the best way to respond when the inevitable happens, and the Big One strikes.

All this may seem pretty dull compared to the sci-fi excitement of quake prediction, but it does at least work. In February 2010, Chile was struck by a gargantuan 8.8 Richter Scale earthquake, one of the most powerful ever recorded. The event wreaked damage across the country and released enough energy to alter the spin of the Earth itself. Yet fewer than 600 were killed – largely as a result of Chile's building codes, which demand that quake resistance be incorporated into homes and offices. In stark and tragic contrast, the Caribbean country of Haiti had been struck by a far weaker 7-Richter event a few weeks earlier. Despite being 500 times weaker, the quake tore apart the tightly packed, badly built shanty towns, killing 220,000.

In essence, the mitigation strategy owes its success to focusing on timescales in which earthquakes are essentially guaranteed – thus eliminating the need for impossibly reliable precursors. There is, however, another strategy which has also proved highly successful. Ironically, it uses the ultimate precursor of any earthquake: the quake itself.

Earthquakes begin when rock can no longer withstand the strain it is being put under and ruptures. The point at which the tearing begins is known as the focus, and it's from here that the seismic waves spread out, wreaking their destruction. These waves come in different forms, however – and, crucially, they don't all travel at the same speed. The fastest are so-called primary or P-waves, back-and-forth motions that travel at incredible speed of around 10,000 to 20,000 kilometres/hour. Then come the secondary or S-waves, up-and-down motions which are far more destructive – but travel at only half the speed. Thus by detecting the P-waves, it's possible to send out an extremely reliable warning of an earthquake 30 to 60 seconds or so before it arrives. This may not sound like much, but it is enough to save lives, as Japanese engineers recognised in the 1960s, while building the famous Shinkansen 'bullet train' network. They installed seismometers that warned train drivers to apply the brakes and reduce the risk of high-speed derailments. By the early 1990s, this had been turned into the Urgent Earthquake Detection and Alarm System (UrEDAS), which identifies P-waves and automatically takes control of endangered trains. It's not infallible: in 2004 a bullet train north of Tokyo was derailed after being struck by S-waves from a 6.8 Richter earthquake whose epicentre was simply too close for anything to be done. Even so, the system confirms the need for extremely reliable 'precursors', even if they can only give a few moments' warning. Astonishingly, there has not been a single quake-related death on the bullet train network in over fifty years of operation and 10 billion passenger journeys across one of the most seismically active parts of the planet.

The UrEDAS system has now been rolled out nationwide, to give at least a few moments' warning of an earthquake. Those indoors can then at least protect themselves by staying away from exterior walls and windows and getting under tables, while anyone outdoors can try to get into open spaces. During the devastating 9.0 earthquake that struck Fukushima in March 2011, a television network carried warnings presaging the arrival of the shock waves by a minute or more, saving many lives. Similar earthquake warning systems have now been rolled out elsewhere, notably in Mexico. A network of seismic detectors has been set up along the Guerrero coast around 350 kilometres from Mexico City, and gives around a minute's warning. Combined with earthquake mitigation, such early warning systems are now succeeding where the sci-fi dream of quake forecasts failed.

The same concepts that ensured its demise hold lessons for the prediction of that most fickle of natural phenomena: the weather. Here progress has undoubtedly been made. According to the UK Meteorological Office, advances in satellite monitoring and computing have led to four-day weather forecasts becoming as reliable as the one-day forecasts of the 1980s. Accuracy figures of around 70 to 80 per cent are claimed for forecasts of sun and rain, and over 90 per cent for temperature ranges. As ever, exactly what 'accuracy' means here isn't clear. In any case, many Brits would struggle to square the figures with their experiences of being caught in unforeseen downpours or bracing for storms that fail to strike.

The problem here is less the 'unreliability' of weather forecasts, and more our failure to know how to react to them. Imagine, for example, you're planning to spend your lunch hour in the park, only to hear that rain has been forecast. Knowing that forecasts of rain are around 80 per cent accurate, it seems obvious you should at least take an umbrella. Yet this ignores the fact that accuracy comes in two forms: correctly predicting something that is true, and correctly ignoring something that is false. In the case of rain forecasts, let's assume that the 80 per cent figure is both the true positive and true negative rates. Then we know that out of 100 times when it actually does rain, the forecast will be right in 80 instances, while out of 100 times when it stays dry, the forecast will be right 80 times. To know how to respond to the forecast, however, we still need one more figure: the chances that it will rain during our lunch break. For the UK, the probability of rain in any given hour is roughly 10 per cent. So now we have everything we need to work out what the forecast really means. And it's not at all what one might expect.

The easiest way to see this is to work out what would happen in 100 cases of our situation: an hour spent outside when rain is forecast. We've learned that of these 100, rain can normally be expected to fall in around 10 while it remains dry in the other 90. The 80 per cent true positive rate of the forecast means that, of the 10 wet lunchtimes, the forecast will correctly predict rain in 8. But that's not the only time rain will be predicted. The 80 per cent true negative rate means that rain will be incorrectly forecast in 20 per cent of cases when it doesn't fall. That's 20 per cent of the 90 dry hours, implying the forecast will wrongly predict rain on a further 18 occasions. So in total there will be 8 + 18 = 26 predictions of rain, of which just 8 will be true positives – a strike-rate of 8/26 or 31 per cent. That's way below what we'd expect if we simply took the Met Office claim of 80 per cent accurate forecasts at face value. But it shows the crucial importance of factoring in the plausibility of any forecast – in this case, determined by the low 10 per cent risk of rain.

We're still left with a decision, however: do we go for our walk, or take an umbrella, or forget the whole thing? Common sense suggests this depends on the consequences of ignoring the forecast, but there's a bit more to it. Just as forecasts can be inaccurate in two different ways, our response to a forecast can also be wrong in two different ways. For weather forecasts, for example, we can either wrongly ignore a forecast that proves right, or trust one that proves wrong. What constitutes the best decision turns out to depend on a surprisingly complex interplay between the prevalence of the event itself, the reliability of the forecast – and our own view of the consequences of making the wrong decision. In other words, what may be the right reaction to an 'accurate' forecast for one person may be wrong for someone else. For example, cranking through the maths,[2] the above figures imply that you should ignore the forecast of rain – unless you think getting rained on is *at least* twice as bad as the frustration of cancelling your walk only to find rain didn't fall after all. What about taking an umbrella? It turns out that you shouldn't bother unless you consider that getting caught without one is *at least* twice as irksome as taking an umbrella, then finding out you didn't need it.

After all this, it's no wonder forecasts have a bad reputation. Even when fairly reliable precursors exist, the forecast itself can still prove worse than useless – simply because it's trying to predict something (often mercifully) uncommon. Most of us also have a flawed understanding of the concept of 'accuracy', and make suboptimal choices in the light of a prediction. Even so, we feel justified in blaming the forecasters when things don't turn out as expected. And hanging over everything is the most basic fact of all: we are dealing with uncertainty and probabilities. As such, the benefits of trusting proven forecasting methods emerge only over time, not every time.

↑UPSHOT

The dream of forecasting natural events is as old as history, but our ability to do so is constrained by fundamental limits. Knowing what they are, and how a forecasting method deals with them, is key to making optimal decisions about future events.

The miraculous formula
of Reverend Bayes

When the US Coast Guard was alerted to what had happened off Long Island one night in July 2013, it seemed clear that a search-and-rescue mission would prove a tragic misnomer. A lobster boat had reported that one of its crew had vanished from the boat around forty miles out in the Atlantic.[1] Somehow he had fallen overboard during the night. Worse still, as he had been working alone, no one knew exactly when or where the accident had happened. When helicopter pilot Lieutenant Mike Deal and his colleagues took off they knew that the chances of spotting one person floating somewhere in over 4,000 square kilometres of ocean were very small. But, crucially, they weren't zero – and that gave the team hope, because of an incredible bit of kit with the power to dramatically increase the odds of success, known as Sarops. It might sound like some fancy box packed with sensors, electronics and microchips, but Search And Rescue Optimal Planning System is actually an algorithm: a mathematical recipe able to process even vague clues about when and where a mariner got into trouble and combine them with insights about local conditions to narrow down the search area dramatically.

That July morning, the Coast Guard fed Sarops with guesstimates of the likely time and place when the fisherman fell overboard, and it came back with the most promising places to look for him. Armed with its insights, Lieutenant Deal and his colleagues took to their helicopter and began searching. As the hours rolled by, more information emerged about the likely timing of the accident, and Sarops produced updated maps for the helicopter crew. Finally, after seven hours and with the fuel gauge telling them to return to base yet again, Lieutenant Deal's co-pilot called out. He had spotted something. They turned around – and there was the fisherman in the ocean swell, waving frantically.

Given the odds of success, what Sarops allowed Lieutenant Deal and his colleagues to achieve that day seems little short of miraculous. But then, they do make use of ideas first explored by a clergyman. Exactly what prompted the English Presbyterian minister and amateur mathematician Thomas Bayes (1702–1761) to develop the formula that bears his name isn't clear. What is beyond doubt is that the outcome has become one of the most controversial results in the theory of chance, whose simplicity and intuitive appeal belie its astonishing power.[2] A clue to the reason for all the fuss can be found in a plain English statement of what Bayes's work implies:

The Reverend Bayes's miraculous rule

New level of belief about something = old level of belief + weight of new evidence

For something supposedly based on the laws of chance, this is a very odd statement, as there's nothing in it about probabilities, frequencies or randomness. Instead, it's all about the altogether more touchy-feely concepts of belief and evidence. And that highlights a feature of probability that Bayes recognised, but which remains controversial to this day: that it can be used to capture degrees of belief. Up until now, we've focused almost entirely on probability in its familiar role in understanding chance events like dice-throwing. But as we noted in the chapter on casinos, this is actually just one form of the beast: so-called 'aleatory' probability (from the Latin meaning 'dice-player'). Bayes's work revealed a far more potent use for the concept: as a way of capturing uncertainty caused not by randomness but by simple lack of knowledge. This 'epistemic' uncertainty (from the Greek word for 'knowledge') is fundamentally different because, in principle at least, we can reduce it using evidence. The question of how, and by how much, is the focus of Bayes's work. As such, it has a direct bearing on the quest at the heart of the whole scientific enterprise: turning evidence into reliable knowledge. Not that you'd guess this from the title of Bayes's work on the subject. Published in 1764, his *Essay Towards Solving A Problem in the Doctrine of Chances* sounds dull, and certainly looks it. Written in archaic English and packed with old-fashioned algebra, it's hard on the eyes, let alone the brain.[3] Add the fact that Bayes himself never got around to publishing it, and it's another miracle we know anything about it at all. For that we owe a debt to Bayes's friend and fellow amateur mathematician Richard Price, who found it among Bayes's papers following his death in 1761. Recognising the essay's implications, Price brought it to the attention of the Royal Society of England, the world's leading scientific academy. The Society duly published it, along with an introduction by Price, who was determined to ensure its significance was recognised. He stressed that Bayes had tackled a problem which was 'by no means merely a curious speculation in the doctrine of chances', but instead one that has a direct bearing on 'all our reasonings concerning past facts, and what is likely to be hereafter'.

Quite why Bayes himself never published the essay isn't known. Perhaps he felt there was far more work to be done, but lacked the mathematical firepower to do it. It's unlikely he'd have guessed his notes would prompt a controversy that would still be raging over 250 years later, with some researchers avoiding using the term 'Bayesian' in their academic papers for fear of sparking rows.

Astute readers may suspect that the source of the trouble lies in the ingredients required by Bayes's rule for updating beliefs. We will see the truth of this shortly, but it helps to understand their origins. They come from Bayes's attempt to answer a perfectly reasonable question left unanswered by the brilliant mathematicians who had founded probability theory in the late seventeenth century. They'd devised formulas that gave the probability of various classic random events – for example, of rolling three 3s in ten throws of a die. Such formulas were of obvious value to gamblers, who could use them to decide whether a bet was worth taking. All they had to do was feed the formula[4] with three numbers: the chances of getting the event in question in any given throw (in this case, 1/6), the number of successes they had to achieve (3) in the given total number of attempts (10), and out would pop the answer (around 1 in 6.5). If the gambler could find someone willing to offer *longer* odds of the event occurring than this – say 10 to 1 against – the gamble was worth taking, as it meant the person offering the odds thought the event less likely than it really is – possibly because of ignorance of how to do the sums. The gambler had to be careful, though, as the same formula could be used by clever bookies to offer odds for deceptively similar-sounding bets with radically different odds – such as the chances of getting *at least* three 3s (about 1 in 4.5), or *no more than* three 3s – which, at 93 per cent, is virtually certain. The formula could cope with all this, while highlighting the fact that precise wording is critical in probability – which as we'll see has generated huge controversy concerning Bayes's work.

On the face of it, Bayes's goal was perfectly straightforward. He wanted to take the usual formulas and flip them around. That is, instead of starting out with a knowledge of what, say, a die can do and then calculating the chances of different outcomes, Bayes wanted to begin with the outcomes, and then work backwards to see what they revealed about the die. Clearly, a formula for that would also be useful to gamblers – not least for detecting cheating. After seeing someone roll four 6s in five attempts, we might suspect cheating, but how could we use the evidence to quantify our suspicions?

In his *Essay*, Bayes laid down the theory for doing such a calculation. He began by proving a neat little recipe for dealing with a very common question: how do we work out the probability of events whose appearance is influenced by prior events? For example, if we've just drawn an ace from a pack of cards and don't replace it

How Bayes's Theorem turns information into insight

The most basic form of Bayes's Theorem shows how the chances of one event A taking place affects those of a subsequent event, B. Specifically, the 'conditional probability' of B, given A is given by:

$$Pr(B \mid A) = Pr(A \mid B) \times Pr(B)/Pr(A)$$

This allows fresh information to be turned into insight. For example, if someone randomly plucks a card from a deck, we know without looking that the chances it's a diamond are 1 in 4. But if we're told the card is red, Bayes's Theorem shows the chances of it being a diamond leap to ½. That's because $Pr(red \mid diamond) = 1$ (as all diamonds are red), $Pr(diamond) = ¼$ and $Pr(red) = ½$, so by Bayes's Theorem $Pr(diamond \mid red) = ½$. Of course, we don't really need Bayes's Theorem to work this out, as everyone knows half the red cards in a deck are diamonds. The point is that the same basic idea works with far more complex problems – such as medical diagnoses.

But there's another simple but critical point worth noting: the danger of blithely switching conditional probabilities around: $Pr(B \mid A)$ may look similar to $Pr(A \mid B)$, but the theorem shows it's only identical if $Pr(B)$ also equals $Pr(A)$. As we'll see, that's crucial to understanding a major scandal that has blighted science for decades.

in the pack, that clearly affects our chances of drawing another ace. Bayes worked out the necessary formula (see box above).

Bayes then went on to show how this simple formula could serve as the foundation of a way of turning observations into insights. For instance, anyone witnessing a coin coming up heads in an unusually high proportion of tosses could use his formula and turn these observations into insights about the fairness of the coin, specifically the chances of the hit-rate truly being around 50 per cent. But as Price stressed in his introduction to his late friend's *Essay*, Bayes had laid the foundation for far more than its boring title suggested: he had opened the way to tackling the *general* problem of turning observations into insight. The link comes from the way in which probabilities can be used to gauge levels of belief. We routinely make the connection in everyday conversation: we talk of believing something being 'odds-on' to occur, of dealing with a '50:50 call', or being '99 per cent certain' about a fact. What Bayes did was to show it's not just possible to quantify our beliefs as probabilities or their close cousins odds, we can also apply the laws of probability to them as well.

Though he never stated it like this, his eponymous theorem can be recast in a form that allows us to update our beliefs in the light of fresh evidence (see box opposite).[5]

In plain English, Bayes's Theorem shows that we can capture our level of belief in some hypothesis or claim using the language of probability. The theorem takes its most simple form when *odds* are used to express the degree of belief that a theory will prove correct or not. Plausible theories – such as the idea that the sun will rise tomorrow – have their plausibility captured by high probabilities and thus 'short' odds, while implausible claims (say, that Elvis lives on the dark side of the moon) have low probabilities and long odds. Bayes's Theorem then shows that we can update our initial ('prior') level of belief in the light of new evidence by multiplying by a factor known as the Likelihood Ratio. This captures the weight of evidence provided by, for example, a lab experiment or a long-term study of lots of people. While the LR might look complex, it too is intuitive. For example, if the probability of getting the evidence we saw, *given* our belief is correct, is very high, the figure above the dividing line will be close to 1, the highest value any probability can take.

So the LR reflects the fact that evidence consistent with our

Bayes's Theorem: how to update your beliefs with evidence

Bayes's Theorem shows how the chances of your belief in a specific claim or theory should be changed in the light of fresh evidence. The simplest form of the theorem measures the impact of the evidence on the *odds* of the claim proving correct:

Odds(Your belief is correct, given new evidence) = LR ×
Odds(Your belief is correct)

where the LR is the so-called 'Likelihood Ratio'. This is what captures the strength of the evidence you've uncovered, and it's determined by taking the ratio of two so-called 'conditional probabilities' – that is, probabilities that are conditional (i.e. depend on) two competing assumptions:

$$LR = \frac{\text{Pr(The evidence being observed, assuming your belief is correct)}}{\text{Pr(The evidence being observed, assuming your belief is NOT correct)}}$$

While this looks complicated it's actually pretty intuitive and easy to use once it's explained (honestly); see text and examples below.

belief carries more weight than evidence that's irrelevant to our belief, or indeed contrary to it (as reflected by a probability of less than 0.5). If, in addition, there's only a *low* chance of getting the evidence we saw, given our belief was *wrong*, that means the evidence is doing a good job of discriminating between our belief and other possibilities. And again, as common sense suggests, that boosts the weight of evidence supporting our belief by increasing the overall value of the LR.

To take an example, if our belief that, say, a patient has breast cancer is just 5 per cent before the screening results came in, we'd set the initial level of belief as odds of cancer of 0.05. Suppose, then, that the screening is done by a method for which the probability of getting a positive result, assuming there is breast cancer, is 80 per cent, while the chances of getting a positive, assuming there isn't breast cancer present (the 'false alarm rate'), is 20 per cent. Then we find that the screening test has a Likelihood Ratio of 0.8/0.2 = 4, and Bayes's Theorem tells us that a positive screening result should increase our initial level of belief that the patient has cancer by 4 times the initial odds of 0.05, giving updated odds of 0.2. Translated back into probability, those odds imply a 17 per cent chance – and thus an 83 per cent chance of the patient still being free of the disease, despite the positive result. That result may well ring a bell: that's because it's exactly what we got using simple proportions and common sense in Chapter 18. And that highlights a key fact about Bayes: in situations where all the necessary information is well defined and measurable, and the range of possible outcomes is pretty simple (such as cancer/cancer-free), there's nothing remotely controversial about Bayes's Theorem.

But as Bayes himself recognised, in many potential uses of his handy theorem, things aren't as straightforward. And those astute readers may have already spotted why. Bayes's rule shows how to update beliefs in the light of new evidence – but clearly that requires some prior belief to update in the first place. In the case of the cancer screening, we can get our prior level of belief about the chances of someone having breast cancer from large studies of the population as a whole. The US Coast Guard searching for the lost fisherman clearly faced a trickier problem. First, the belief wasn't a simple true/false dichotomy; it was about which areas were best to search. Then there was the problem that the search teams had only some vague prior belief of where the accident may have happened. But they were then able to benefit from the key feature of Bayes's Theorem: its ability to continually update beliefs. Thus when the initial guesses about the fisherman's whereabouts proved incorrect, that – plus new insights about currents and the like – could be fed in to become the new initial level of belief, which was itself updated in a series of iterations which homed in on the target.

Yet at least the Coast Guard had some vague idea of where to start. Clearly, there was no point looking in the Pacific, say. But what do we do if there's no good evidence at all? Imagine that we're taking part in some new casino game, and we suspect it might be rigged. How should we capture our initial ('prior') belief that the game is unfair, given that we've nothing really to go on? Bayes himself suggested a way of tackling this issue – dubbed unimaginatively the 'Problem of Priors' – by using the observations at hand to give a first guess. It worked – but only in certain circumstances. That led to his formula being seen as having only restricted practical value,[6] and not even Price's efforts to promote the work of his friend, and its publication by the world's most famous scientific institution, could stop it being almost entirely ignored.

Fortunately – and as so often with major breakthroughs – Bayes had not been alone in pondering how to turn observational data into insight. As one of France's most brilliant exponents of the application of mathematics to real-life problems, Pierre Simon de Laplace had been pondering the same issues for years when, in 1781, he learned of Bayes's work via a colleague. He too had been wrestling with the 'Problem of Priors', and hit upon an apparently obvious solution: if we have no initial insight about, say, the chances of a specific coin giving heads, why not just assume it was equally likely to lie at any value between zero and 100 per cent? Known variously as the 'Principle of Insufficient Reason' or 'Principle of Indifference', it was simple to apply, and seemed to open up a host of applications. Laplace himself set about using the formula to attack problems in everything from demographics and medicine to astronomy. By the time of his death in 1827, he had given Bayes's rule both its modern format and the imprimatur of authority (indeed, a strong case can be made for calling it the Bayes–Laplace Theorem). But soon his methods were coming under attack from a new generation of researchers, who homed in on what they deemed the Achilles heel of the whole process: the problem of setting prior levels of belief in the absence of any evidence. Some objected to Laplace's use of 'indifference' as the starting point for the calculations; others disliked the use of probabilities as seemingly vague 'degrees of belief', rather than as nice, concrete frequencies of events.

The most vociferous criticism came from those who saw the Bayes–Laplace Theorem as a threat to the whole scientific enterprise. For them, the theorem's need for some statement of initial belief threatened the most cherished aspect of scientific research: its objectivity. For in the absence of any prior insight, what was to stop researchers applying the theorem using prior beliefs based on guesswork – or worse still, their own subjective opinion? That in turn would allow researchers to take observational data and draw any conclusion they liked, simply by tweaking the prior level of belief to get the result they wanted. What self-respecting scientist could stand by and allow such practices to worm their way into the dispassionate quest for Truth?

By the 1920s Bayes's Theorem had been excommunicated from the scientific enterprise. While the most influential statisticians of the day accepted Bayes's neat little recipe for working out 'conditional probabilities' of events affected by others, they rejected its role in turning evidence into insight. Instead, they devised a whole new toolkit based on apparently wholly objective 'frequentist' concepts, where probabilities remained just the frequencies with which outcomes occurred when given the opportunity. In essence, these attempted to avoid the Problem of Priors raised by Bayes's Theorem by sticking to the original formulas of probability theory, which simply give the outcomes one can expect, *assuming* the cause is already known. So, for example, the frequentists would set about investigating whether a coin was fair by *assuming* it was, and then using the standard formulas of probability to work out what should then be observed, *if* this assumption were true. If it turned out that what was observed had only a very low chance of appearing if the coin was fair, frequentists would then argue this was evidence that there was only a small chance of the coin being fair, and thus one should suspect cheating.

If this doesn't quite sound right, congratulations: you've just spotted a flaw in reasoning that many – perhaps even most – research-ers have failed to grasp over the last century or so. The argument commits the fundamental blunder of claiming that the probability of A given B is the same as the probability of B given A. In the example given, the specific error lies in assuming it's OK to argue that

Pr(evidence from tosses, given coin is fair) = Pr(coin is fair, given evidence from tosses)

But as Bayes showed in his completely uncontroversial results concerning conditional probabilities, flipping them around like this is a very dangerous thing do. As we saw with the example in the box on p141, with simple probability questions it leads to results that are just flat wrong – such as the probability that a card is a diamond, given it's red (50 per cent probability), is equal to the probability a card is red, given it's a diamond (which is 100 per cent). When used to turn evidence into insights, however, care-lessly flipping conditional probabilities around like this is a recipe for disaster. That's because it commits the logical fallacy of first assuming something is true to reach a deduction, and then using the deduction to test the assumption.

Bayes's Theorem shows that the only way we can flip condi-tional probabilities around is by bringing in extra information. And in the case of drawing inferences about our beliefs from data, that means including some prior probability that our belief is correct. That, in turn, means we must face up to the Problem of Priors. As we've seen, it's not always a problem: sometimes there's an obvious source of such prior insight – such as past research. But often there isn't, and we must face the fact that drawing insights from data can involve subjective guesswork. Crucially, however, Bayes's Theorem shows that – as the US Coast Guard demonstrated – as evidence accumulates, whatever initial guesses were used tends to become ever less important, and the evidence 'speaks for itself'.[7]

As the frequentist methods became ever more popular, some statisticians repeatedly tried to warn of the dangers of sweeping all this under the carpet. They were more or less completely ignored for decades. Even today, many researchers continue to use frequentist methods to extract insights from data. As a result, countless claims in fields ranging from economics and psychology to medicine and physics are at best questionable, and quite possibly flat wrong. Evidence for this is now starting to rear its head, as researchers struggle to replicate 'discoveries' based on the flawed logic of frequentism. We'll look at this deeply disturbing issue later, but perhaps the most shocking aspect of it is that frequentism's flaws have been tolerated for so long. That's changing. What are now known as Bayesian methods are increasingly being put to use by researchers in a host of fields. Part of the reason is their power. Until recently, the full armamentarium of Bayesian tools wasn't available to researchers who – like Bayes himself – found themselves facing problems doing the sums needed to apply them to real-life problems. That's now been solved by the emergence of cheap, plentiful computing power, which allows Bayesian methods to be used in very sophisticated problems involving lots of competing theories.

At the same time, researchers are becoming increasingly aware of the many virtues of the Reverend Bayes's miraculous formula. And as we'll see shortly, one doesn't even need to plug in any numbers to benefit from them.

↑UPSHOT

The laws of probabilities don't just apply to trivial chance events like coin-tosses. They can also be used to capture the otherwise fuzzy notions of belief and evidence, and combine them to produce fresh insights. Key to the process is Bayes's Theorem, long controversial but increasingly seen as the best way of making sense of evidence.

When Dr Turing met Reverend Bayes

I n April 2012, the UK government's communications headquar-
ters, GCHQ, finally revealed one of its most closely guarded
secrets. It took the form of a declassified 44-page technical document outlining an astonishingly powerful method for breaking
enemy codes. Some insight into just how powerful can be gauged
by the fact that it had been written during World War II, but it
had taken over seventy years before, as one GCHQ insider put
it, mathematicians had finally 'squeezed the juice' out of it. The
release of so secret a document was notable enough, but so was its
author: Dr Alan Turing, the brilliant Cambridge mathematician
who played a now-celebrated role in breaking Nazi codes, and went
on to pioneer the creation of computers.

The media naturally made much of Turing's authorship of *The
Applications of Probability to Cryptography*. For the cognoscenti,
however, there was something even more impressive about the
document, with its talk of evidence, probabilities and use of prior
evidence. It was final confirmation of what had been suspected
since at least the 1970s: that Turing and his colleagues at the Allied
code-breaking centre of Bletchley Park had made extensive use of
Bayes's Theorem. The first hints of its central role emerged in a
paper on the wartime statistical work of Turing published in 1979
by fellow Bletchley mathematician and Bayes enthusiast I. J. 'Jack'
Good.[1] Even then the mere mention of Bayesian ideas was enough
to provoke furious denunciations of the use of prior beliefs from

leading statisticians. Yet now it's clear that while leading statisticians were waging an intellectual war against Bayes and all his works in the outside world, Turing and his colleagues at GCHQ had been using them in great secrecy to bring victory in the all too real World War II and the conflicts that followed.

As students, Turing and Good had managed to dodge the frequentist dogma then sweeping through the research community. They turned to Bayes's Theorem simply because it seemed ideal for the core challenge of code-breaking: turning clues and hunches into insights. Using it, they worked backwards from the observed data – intercepted enemy signals – to work out the most likely settings for the encryption devices like the Enigma machine, used by the Nazi armed forces to encrypt their operational communications. Its gears and wiring could scramble messages in 15 billion billion different ways, prompting the head of Bletchley Park to express his doubts that its messages could ever be 'broken'. He had reckoned without the power of Bayes to take even faint clues, combine them with data, and repeat the process over and over until the right settings emerged – and the messages became readable. The Bletchley code-breakers went on to turbocharge Bayes with the first primitive electronic computer, Colossus, and used the combination to break the far more demanding Lorenz machine, which Hitler himself used for his most secret communications with field commanders. After the war, it's likely Bayes was put to work in the West's biggest Cold War triumph: Project VENONA. A wartime blunder by Soviet intelligence introduced a tiny flaw into the code system used by its top spies. Exactly how this was exploited remains classified to this day, but there is evidence to suggest Bayes and computers again played a role. By the time Project VENONA was ended in 1980, it had unmasked many of the most famous spies of the Cold War, among them Klaus Fuchs, Alger Hiss and Kim Philby.

After rejoining academia in the 1960s, Good became one of a handful of torchbearers for Bayesian methods during their long sojourn in obscurity. Sometimes he would have to sit through lectures dismissing Bayesian methods, prevented by secrecy rules from drawing on his own experiences to rebut them.[2] Such intense secrecy was well founded. In 1951 two mathematicians working on VENONA for US intelligence were allowed to published a paper incorporating Bayesian ideas – and its value was immediately noticed by Soviet code-breakers.[3]

They would have relished getting hold of Turing's top-secret document, which provides a primer on applying Bayes's Theorem to the general problem of code-breaking. Starting from basic principles, Turing applies the theorem to increasingly complex systems, giving worked examples as he goes. Most of it is, inevitably, pretty recondite, but there are two features of Turing's use of Bayes's Theorem that hold lessons far beyond the secret world of the code-breaker. The first is Turing's insouciance when faced with the supposedly critical 'Problem of Priors' – that is, the setting of an initial level of belief from which to begin the Bayesian process of updating using fresh evidence. He had no qualms about resolving the problem with a judicious mix of hard facts and educated guesswork. Such practices were regarded as anathema by the most influential academic statisticians of the day (and still prompt controversy today). Fortunately for the Allies, their influence did not extend to Bletchley. Even if they had, it's unlikely they would have deterred Turing, who was already renowned for his pragmatism and disregard for authority. As he showed, as long as the initial guesstimates weren't too outlandish, Bayes's Theorem made them progressively irrelevant as fresh evidence came in, resulting in useful insights. Proof of Turing's claim could hardly have been more impressive: the breaking of enemy code systems which hastened the defeat of the Axis powers and saved countless lives. Ironically, the rehabilitation of Bayesian methods after the war might have been speedier had they not been so successful in such vital and thus secret applications.

But there is another aspect of Turing's use of Bayes's Theorem that makes it accessible even to the most non-mathematical. Despite their brilliance, neither Turing nor any of his colleagues relished unnecessarily complex calculations. Like the rest of us, they found addition easier than multiplication, and this led them to recast Bayes's Theorem in a format that is easier to use, even more intuitive than the original format, yet retains all its power.[4] Here it is:

Dr Turing's version of Bayes's Theorem

New level of belief in theory =
old level of belief + Weight of Evidence

where the *Weight of Evidence* is depends on the so-called *Likelihood Ratio* (LR), the ratio of two conditional probabilities: the chances of getting the data observed, assuming one's belief is correct, divided by the chances of doing so if one's belief is wrong. That is:

$$LR = \frac{Pr(data \mid belief \; is \; correct)}{Pr(data \mid belief \; is \; wrong)}$$

The first thing to note about this little recipe is that it now exactly mirrors how we talk about evidence and beliefs. We now have a formula in which data are turned into weight of evidence which adds to our level of belief. In putting it into this format, we have taken the most basic means of capturing our beliefs – as probabilities ranging from 0 via 0.5 to 1 – and turned them into so-called log-odds, which extend from minus infinity at one extreme and then pass through zero to reach plus infinity at the other. Thus the way strength of belief is captured now stretches from implacable scepticism at one extreme, through to utter certainty at the other, via no view either way in the middle – a beautifully symmetric and natural measure of our belief. At the same time, unlike the values of 0 and 1 of probability, the log-odds equivalents of minus and plus infinity serve warning on us of how extreme such levels of belief are. That is, they underline the fact that implacable scepticism and utter certainty have no place in the real world.

Plugged into Bayes's Theorem, they also show the irrationality of holding such extreme levels of belief, as they cannot be altered by any amount of evidence. Thus this formulation of Bayes's Theorem not only captures the seemingly ineffable notion of belief, but also shows how to change it in the light of evidence – while warning us of the futility of aspiring to God-like certainty. The Reverend Bayes himself would surely have approved.

As for how new evidence should affect our beliefs, this becomes more in line with common sense too. As the box opposite shows, we can now add or subtract weight of evidence to our level of belief. Exactly how much we add or subtract is dictated by a simple calculation: we take the chances of observing the data if our theory is *true*, and divide them by the chances of doing so if our theory is *false*. Then, as one would expect, if the evidence is more likely on the assumption the theory is true than if it's false, it *adds* weight of evidence in favour of the theory. On the other hand, if the evidence is more likely on the assumption the theory is false, then that subtracts weight of evidence. But there's also third possibility we mustn't ignore: that the evidence is equally likely to be observed, *regardless* of whether the theory is true or not. As the box opposite shows, this leads to a 'Likelihood Ratio' of 1, which the log-transform trick turns into a weight of evidence of zero. In other words, such evidence makes zero difference to the weight of evidence for or against the theory being true.

All this is now summed up in the intuitive set of rules about gauging weight of evidence for or against a specific belief (see box overleaf).

Now we have what we need to use Bayes's rule to update our level of belief in a theory. And the 'theory' in question doesn't have to be some esoteric explanation of, say, subatomic forces or the origin of the universe (though Bayes can and is used for such things); it can be *any* hypothesis, from whether a particular setting was used to encode a particular Enigma message to whether new evidence supports belief in telepathy. Bayes's Theorem doesn't care: whatever notion we want to assess, it tells us what we need to consider to create weight of evidence for or against, and then what

How to make sense of evidence

To gauge what impact evidence should have on our level of belief in some theory or claim, we need to know (or at least guesstimate) two probabilities: the chances of getting the evidence assuming our theory is right (call these R), and the chances of getting the evidence assuming our theory is wrong (W). Then Bayes's Theorem shows that:

1. If R is **greater** than W then we have **positive** weight of evidence that adds to our belief that the theory is correct.
2. If R is **less** than W – that is, the evidence is *less* likely to emerge if the theory is right than wrong – then the weight of evidence is **negative**, and should weaken our beliefs.
3. If R **equals** W, the evidence is just as likely to have emerged *regardless* of whether the theory is right or not; it supplies **zero** weight to our beliefs, and we should be left unmoved by it.
4. If we don't have, and can't even guess, either R or W then we can't tell whether the evidence is more likely or less likely to have emerged if our theory is right – and we must be wary of reaching any judgement *at all*.

impact that has on our level of belief. Indeed, it amounts to a one-line formula for 'How to make sense of evidence'.

The first two rules in the box above are most useful when we have some hard numbers to plug in. For example, during their code-breaking work, Turing and his colleagues used sophisticated probability theory to estimate the chances of getting the glimpses of readable text they extracted, on the assumption that they'd got the right settings of the enemy code machine, or by fluke alone. They then added this fresh weight of evidence to their existing level of belief that they had the right settings, and continued – homing in on the complete 'break'.

The third and fourth rules, in contrast, are often useful for testing claims even in the absence of hard numbers. Take, for example, the claim that it's possible to discover whether people like butter by holding a buttercup under their chins and looking for a telltale yellow glow. It's a lovely idea, and one long used by parents to keep kids amused on summer days – kids who duly try it out on their friends and uncover its stunning reliability. Yet most grown-ups know there's something not quite right with this apparent support for the test. Bayes's Theorem crystallises those doubts, but goes farther – highlighting the critical rule for testing *any* theory. As we'll see, it's one that routinely catches out many otherwise smart people.

The suspicions about the 'buttercup test' centre on the fact that as most people like butter, the chances of the test coming up positive are very high, even if the test is nonsense. Bayes's Theorem confirms these suspicions. Rule 3 warns us that if the chances of getting the evidence are just as likely *regardless* of whether the claim is true, then it supplies zero weight of evidence. Thus while our kids may be impressed by seeing the yellow glow under the chin of everyone who likes butter (at least, on sunny days), Bayes's Theorem shows that this is merely half the story. The test can only claim to generate real weight of evidence if positive results are not only *more likely* with those who do like butter, but also *less likely* with those who don't – and that demands that we carry out tests on *both* kinds of people. This demand for comparison tests is often missed by adults, let alone kids, and is underlined by Rule 4, which is even easier to use, and even more widely applicable. So, for example, if we hear reports of some amazing new test for a medical condition, we need to know more than whether the test gave positive results for patients with the condition (so-called 'true positives'). For the test to generate useful weight of evidence, comparison tests must also have been done to check for positive results for patients who didn't have the condition ('false positives'). Without that, as Rule 4 states, we must be wary about reaching any judgement about the value of the test at all.

Even if the researchers have done all this, their diagnostic test is useful in adding weight of evidence only to an existing level of belief – and Bayes's Theorem shows that if this was very low (because, say, the condition is very rare), then even after adding the hefty weight of evidence, the updated level of belief might still be pretty low. Of course, Bayes is at its most powerful if we can plug in numbers to get a quantitative answer (as we did in Chapter 20), but the point has been made: we must not be overly impressed by claims based solely on impressive true positive rates: we need more than that.

When we have it, the results can – as Turing and his colleagues showed – change the course of history. Fortunately, we did not need to wait for the release of his report for the power of Bayes to be recognised more widely. Its ability to quantify the central process of science – updating knowledge in the light of fresh evidence – is finding uses in ever more fields. Clinicians testing a new therapy exploit its ability to combine existing knowledge with fresh data, which allows them to reach a decision about efficacy faster, more reliably and using fewer patients.[5] Palaeontologists trying to unravel the evolution of *Homo sapiens* use Bayesian methods to compare rival theories and focus on the most plausible,[6] while cosmologists use them to pin down the properties of the universe with unprecedented precision.[7]

Bayes's Theorem is also being put to a host of less exalted but no less impressive uses, speeding our online searches, fixing our typing errors and protecting us from all those viA*g_rA D#Als through its innate ability to learn from what is already known. And in a wonderful example of history repeating itself, the theorem used by Turing and his colleagues to such triumphant effect during World War II is now being used against a new global enemy: cyber criminals.

From multinational media empires to oil companies, defence contractors to dating sites, computer networks are now under constant attack from hackers. In the cyberspace equivalent of Darwinian evolution, each countermeasure is met with an increasingly sophisticated response – and growing recognition that the old techniques of passwords and encryption are no longer enough. Many attacks are now being perpetrated by insiders able to bypass security systems. Yet there's one thing that never changes about cyber criminals: by definition, they're after sensitive information. No matter how long they pretend otherwise, they eventually have to reveal their true intentions – snooping in personal files, for example, or attempting to download data. In short, like their real-world equivalents, cyber criminals have 'MOs' that can be learned and looked for.

Identifying such activity is now seen as vital in the fight against cyber crime, and leading the charge is a British-based company known as Darktrace. Many of its staff are alumni of GCHQ, the modern-day equivalent of Bletchley Park, where Turing and his colleagues performed their miracles. The driving force behind Darktrace's strategy is a method for learning how computer networks look when they're fine, thus revealing when they're not. And at its heart is none other than the Reverend Bayes's miraculous formula.

⬆UPSHOT

Even in the absence of hard numbers, Bayes's Theorem helps reveal exactly what questions we should ask about evidence. It also alerts us to when we're being told only half of what we need to know – and sometimes even less.

Using Bayes to be a better judge

Around 4 a.m. on 21 July 1996, after hours of questioning by Louisiana detectives, Damon Thibodeaux finally broke down and confessed to the murder of his cousin. Her body had been found the previous day on the banks of the Mississippi, and Thibodeaux did not stint in revealing what he had done: how he had hit her in the face, raped her and finally strangled her with some wire from his car. The trial lasted just three days, and the jury took less than an hour to deliver their verdict: guilty. He was sentenced to death for murder and aggravated rape.

Thibodeaux spent the next fifteen years on Death Row, until finally, in September 2012, he was exonerated on all counts and released. He had become the 300th person to be proved innocent using DNA evidence in the United States – but just the latest of the countless thousands who down the centuries have been convicted on the basis of flimsy evidence.

Following his release, Thibodeaux explained how he had come to believe that he had committed the crime: a mix of sleep deprivation, relentless pressure and an overwhelming desire for it all to just end. Even during the trial it was clear that his 'confession' was based on a mix of clues picked up from the detectives' allegations, and simple invention. The victim had been struck with a blunt instrument, not a hand, strangled with wire from a tree, not from his car – and there was no evidence of sexual activity, forced or otherwise. Thibodeaux even told his interrogators: 'I didn't know that I had done it, but I done it.'

It was, in short, a classic case of false confession, serving only to add weight to the belief that this age-old form of 'evidence' is flimsier than the paper it's scrawled on. No one knows this better than the members of the Innocence Project, set up in 1992 at the Cardozo School of Law in New York to re-examine apparent miscarriages of justice. At the time of writing, their work has led to the exoneration of over 300 people convicted of serious crimes they did not commit, but for which they've typically served well over a decade in jail – many, like Thibodeaux, on Death Row. Over a quarter of the wrongful convictions overturned by the Innocence Project have involved false confessions. The rate in nations with less regard for due process hardly bears thinking about.

Many of us have an innate distrust of confessional evidence – and it is an attitude supported by the most basic implication of Bayes's Theorem. As we saw in the previous chapter, for any source of evidence to add weight to our beliefs about a theory, a very specific condition must apply. And for a confession to add weight of evidence to our belief in a person's guilt, that condition is:

Pr(confession, given guilt) must *exceed* Pr(confession, given innocence)

Put more bluntly, we must be confident that the chances of getting a confession from the guilty are greater than those of getting one from the innocent. One can of course argue about this – and that is precisely the point: it clearly cannot be regarded as beyond question in every case. Indeed, one can imagine being in the situation of Thibodeaux and pressurised beyond endurance until one is willing to say anything – the only question being, just how much would it take. For some, it might take extreme torture; for others the mere possibility of fifteen minutes of fame on TV has proved motivation enough. What Bayes highlights is the constraint put on the two probabilities – and the fact that it is far from guaranteed to hold. In fact, a moment's thought suggests the exact opposite may hold for certain types of crime. For example, if a professional gangland murder has been committed, we can be pretty sure the 'perp' is likely to be, well, a professional gangland murderer. Are such people, with their codes of *omertà*, really more likely to confess during interrogation than someone innocent brought in for questioning? Remember that Bayes shows it's not enough for such people to be *as likely* to crack; they have to be *more* likely to crack for confessions to be useful sources of evidence of guilt in such cases.

Such doubts are even stronger with terrorist-related crimes, where it's known that perps are often specifically trained to resist interrogation. So now we have a situation where those guilty of terrorist outrages are in fact *less* likely to crack under interrogation than an innocent person. In this case, Bayes tells us something rather shocking: that the very fact that someone accused of a terrorist outrage has confessed makes it *less* likely they are the perps. Bayes is thus telling us that those convicted of such crimes on the basis of their confessions may well prove to be victims of miscarriages of justice. It may not be a coincidence that confessional evidence from supposed terrorists looms large in some of the most egregious miscarriages in many countries, such as the Guildford Four and Birmingham Six terrorism cases in the UK in the 1970s.[1]

Clearly, in many cases there is more compelling evidence than just a confession, provided by additional, and more reliable, sources of evidential weight – such as that based on forensic science. Or at least, it would be nice to think so. The trouble is that all too many forensic science tests have been accepted in court without undergoing the 'Bayesian tyre-kick' to check they actually do add weight of evidence.

Take the notorious Birmingham Six case of 1975, where six men were convicted for the IRA attack on two Birmingham pubs, which killed 21 people and injured over 180. Four of the six men signed confessions soon after being arrested, but it wasn't only confessional evidence that sealed their fate. Three of the men also tested positive in the so-called Greiss Test for contact with explosives. According to the forensic scientist, the result was so strong he was '99 per cent certain' that some of the defendants had handled explosives.

Quite what he meant by that isn't clear; most likely he was referring to the fact that the test is very effective at detecting traces of nitrites, found in nitroglycerine. The trouble is, as Bayes's Theorem shows, knowing that a source of evidence has a high probability of coming up positive in the right circumstances (that is, having a high 'true positive' rate) is only half the story; to establish its weight of evidence we also need the false positive rate – and, moreover, this needs to be *lower* than the true positive rate. This key fact was never raised at the trial. Astonishingly, it was only in 1986 – over a decade after the convictions – that UK government forensic scientists put together a report on this issue. They found that the Greiss Test was quite capable of giving positive results when applied to the hands of people who had simply played cards, or not washed their hands after urinating. In other words, the test had an impressive true positive rate, but also a significant false positive rate, undermining its evidential weight.[2] Nor was this the first time doubts had been raised about such tests: the same issue had been identified a decade before the Birmingham Six even came to trial, with false positives being generated by a very similar test that had been used in the USA since the 1930s.

Far more worrying, however, is the fact that there are still forensic tests in use that have never undergone a proper Bayesian tyre-kick, resulting in innocent people being incarcerated. According to the Innocence Project, almost half of the 300-plus cases of miscarriages of justice it has uncovered involve forensic tests that have been misinterpreted, bungled or never properly validated. Even such well-known and widely used techniques as hair microscopy, shoe print analysis and bite mark comparisons have never been put through the Bayesian wringer to assess what weight of evidence – if any – they provide. In contrast, the Innocence Project has plenty of evidence of their failings – such as the case of Steven Barnes, convicted of the rape and murder of a woman in Whitestown, New York, in 1989. To the jury, this must have seemed like a no-brainer case. While the eyewitness evidence was so-so, the forensic evidence was compelling. Soil on Barnes's truck tyres had similar character-istics to that at the crime scene, and the pattern of fabric from the victim's jeans shared features with an imprint found on the truck. Perhaps most telling of all, microscopic examination of two hairs found in the truck had traits different from those of Barnes – but again, similar to those from his alleged victim. Other tests proved inconclusive, but the jury had heard enough, and Barnes was sentenced to a minimum of 25 years. Barnes became one of the Innocence Project's first cases, and its team identified a host of flaws in the case against him. And among them was the fact that the hair, fabric matching and soil tests had never been scientifically validated.

Barnes was finally exonerated in 2009, almost twenty years after his conviction. That same year, forensic science found itself in the dock of the court of scientific respectability, with no less than the US National Academy of Science leading the prosecution. In a report entitled *Strengthening Forensic Science in the United States*, the NAS did not mince its words: the data demanded by Bayes's Theorem to establish evidential weight 'are key components of the mission of forensic science' and explicit and precise statements are 'absolutely critical'.

Fortunately for the likes of Barnes, Thibodeaux and many other innocent people, there is one forensic test with both a solid scien-tific basis and well-established true and false positive rates: DNA profiling. Since it was first used in 1987 (as it happens, to exon-erate someone who had falsely confessed to a double murder in England), it has not only helped catch countless criminals, but also revealed the failings of many supposedly 'scientific' forensic tests. DNA profiling has become the gold standard to which the Innocence Project and countless others have turned in the search for the truth. Yet Bayes's Theorem shows that even DNA profiling can be undermined by failure to understand the process by which evidence becomes insight.

DNA profiling owes its reputation to the fact that everyone apart from identical twins has a unique genetic profile, packaged up in the famous double helix molecule crammed into their cells.

This gives the technique an extremely high true positive rate: it's virtually certain that the DNA found at at a crime scene will match those who were there – including the guilty person. It thus has a virtually 100 per cent true positive rate. But as ever, Bayes warns us not to be overly impressed by this, and to demand to know the chances of getting a match from someone who wasn't at the crime scene – the false positive rate. Bayes's Theorem then shows that the greater the difference between the true and false positive rates, the higher the weight of evidence. The exact figure depends on the quality of the DNA, and how many 'matches' are found with the sample taken from the suspect. Put simply, because of the chemical nature of DNA, it's not uncommon to get many matches from a sample, driving the false positive rate down to as low as one in several million. So now we have both components needed for the weight of evidence, and DNA profiles clearly provide huge amounts of it. Plugging the figures into Bayes's Theorem shows that the technique can increase the prior odds on guilt by a factor of several million. But it also makes clear that we still need to know what those prior odds are before we can reach any conclusion about the guilt or innocence of the accused. And if there's very little other evidence, that prior level might be extremely low. For example, if all we knew before the DNA test is that, say, the criminal was a man from England, the prior level of belief that the suspect is guilty is just 1 in 30 million – the male population of England. So even after being amplified by a factor of several million, we can still end up with a level of belief in guilt of around 1 in 10 – which is far from 'beyond all reasonable doubt'. Despite all this, and the clear and present danger of misinterpretation, DNA evidence is still routinely presented without any reference to Bayes, which makes clear exactly what the DNA evidence means, and how to incorporate it with other evidence to reach a final verdict. Juries can – and do – find themselves trying to make sense of statements by forensic scientists such as 'The probability of getting as good a DNA match from someone unconnected to the crime scene is 1 in 3 million'. Without Bayes on hand to make clear that this is just a statement of the false positive rate, there is a high risk of juries confusing it with the chances of the suspect being innocent, which at just 1 in 3 million seemingly implies guilt far beyond reasonable doubt.

Given the central role of evidence in the courts, and of Bayes's Theorem in making sense of it, there is clearly a need for anyone dealing with forensic evidence to have some idea of its implications. Simply being aware of the rules governing weight of evidence is enough to avoid the most egregious pitfalls in assessing evidence. Yet, incredibly, in the UK the judiciary has specifically rejected this modest proposal. In a 1997 ruling widely condemned at the time and still the cause of much debate, the English Court of Appeal ruled that it is 'not appropriate' for juries to make sense of evidence using 'mathematical formulae such as the Bayes Theorem [sic]', as this 'would encroach upon the jury's task of weighing up all the evidence together'. Indeed it would, but in a way likely to lead to less reliability placed on flawed evidence, less confusion over what it means, and fewer miscarriages of justice.

↑UPSHOT

It's comforting to think that juries no longer reach verdicts on the basis of trial by ordeal, hearsay and false confessions. Yet the weight of evidence provided by many supposedly 'scientific' forensic tests has never been properly established. Until it is, they will continue to play a role in egregious miscarriages of justice.

A scandal of significance

As academic journals go, *Basic and Applied Social Psychology* isn't a big hitter. Founded in 1980, it has a specialist readership, a modest circulation and nothing like the influence of front-rank research journals like *Science* or *Nature*. Even so, in 2015 the *BASP* managed to spark controversy in scientific circles when its editors declared they would no longer accept research claims backed by 'significance tests'.

This sounds like one of those issues only academics understand or care about. Yet it should concern us all, because the *BASP*'s editors had highlighted an issue that threatens the reliability of scientific research. It centres on the methods widely used by researchers to decide whether they've found something worth taking seriously. Like a kind of quantitative litmus test, these methods are applied to experimental findings to find out whether they can be deemed 'statistically significant'. This is a matter of critical importance, as such findings have a much higher chance of being published by research journals, bringing kudos and funding to the researchers. In some cases they can spawn whole new areas of research, influence public policy – and even change global practices.

The trouble is, there are some very serious problems with this litmus test. First, the criterion used to decide statistical significance is unreliable, being scarily prone to passing off fluke results as genuine effects. Secondly, it's misleading, encouraging those who use it to believe what they've found is truly 'significant', in the sense of important. But most worrying of all, many – perhaps even most – researchers don't really understand how or why their results have passed the test of statistical significance. As a result, a substantial fraction of the countless research claims made over the decades on the basis of 'statistical significance' are meaningless nonsense.

The very idea that generations of scientists have being using a fundamentally flawed technique for making sense of evidence sounds outrageous. Were it true, surely it would have been pointed out decades ago? And if the blunder really was that serious, surely there would be a plethora of evidence that many research findings are undermining scientific progress? In fact, they have, and there is. Ever since they were first adopted for making sense of scientific evidence over eighty years ago, significance tests have come under attack by some of the most eminent statisticians of the day.[1] Even their inventor, Professor Ronald Fisher of Cambridge University – widely regarded as one of the founders of modern statistical methods – expressed concerns over their misinterpretation. Every so often, academic journals and learned societies have taken up the issue, pondered it for a while – only to drop it again. The *BASP*'s ban on significance testing briefly garnered headlines in the academic media, but it too looks unlikely to bring about more widespread change.

Such apparent complacency is even harder to understand given the plethora of evidence that significance testing isn't fit for purpose. For years, the evidence was largely anecdotal; much of it in the form of studies of health issues which never seemed to reach any kind of consensus, as one might reasonably expect if there was some genuine effect at work. Mobile phones and brain cancer, overhead power lines and childhood leukaemia, genetic links for all kinds of traits – the evidence ebbed and flowed, without any sign of resolution. Studies sometimes contradicted each other with almost risible alacrity, one leading journal publishing a headline-grabbing research finding one week, only for it to be apparently debunked soon afterwards.[2]

There's no shortage of explanations for such failures to reach a consensus. As we've seen in Chapters 10 and 11, studies can be undermined by a host of factors, such as a lack of randomisation. They can be too small to detect a real effect, or so large that researchers have a high chance of finding some impressive result by fluke alone – if they just keep looking hard enough.[3] All this has provided convenient camouflage for an inconvenient truth: that 'significance testing' can make the drossiest of data seem like scientific gold.

The evidence has been around for decades for those willing to see. In 1995, the leading research journal *Science* carried a special report[4] on what might be called the 'Curious Case of the Vanishing Breakthrough'. The focus of the report was epidemiology, a field in which researchers routinely make headlines with claims that some or other activity, from drinking coffee to using aluminium saucepans, is linked to some or other health effect, from heart attacks to Alzheimer's disease. Statisticians interviewed for the *Science* report warned that the whole field was vulnerable to the widespread misconception of what statistical significance really means. Yet their concerns came across as academic nitpicking compared to the host of other, more familiar causes of unreliable conclusions, such as inadequate sample sizes and poorly chosen study groups. Even so, the strange fade-out of supposedly impressive evidence continued across the research agenda, from psychology to nutrition to economics. A decade later, the distinguished medical statistician John Ioannidis of Stanford University published a celebrated paper with the title 'Why most published research findings are false',[5] arguing what many statisticians had been saying for decades: that statistical significance testing is a 'convenient, yet ill-founded strategy' for reaching scientific conclusions. His implication that over 50 per cent of *all* research findings are wrong can be criticised as unsubstantiated at best, and probably a gross exaggeration. That said, attempts to gauge the scale of the problem by replicating published studies suggest that around one in five research claims are false positives, and higher still in some disciplines.[6] Given the colossal amount of time, effort and money (currently around $1.5 trillion a year globally[7]) spent on scientific research, if these figures are even remotely correct, they represent a scandal of stunning proportions.

So what exactly is wrong with these techniques, devised and promoted by one of the founders of modern statistics, and still taught and relied on by researchers worldwide? Why are researchers so reluctant to abandon them – and what should they be doing instead? By now, it may come as little surprise to learn that the answers lie in Bayes's 250-year-old recipe for making sense of evidence – and the issues scientists have had with its implications ever since.

The fundamental flaw in the way scientific evidence is usually assessed lies in this simple fact: as Bayes showed, you can't blithely take statements like 'the probability of A, given B', flip them around to give 'the probability of B, given A' and assume the answer must always to be the same. Sure, it *can* be – if the events A and B are independent. For example, if we're using a fair coin, there's clearly no problem assuming that because

Pr(we get heads on second toss, given we got tails on the first) = ½

we can just flip the events around to say that

Pr(we get tails on second toss, given we got heads on the first)

also equals ½, because the two events are independent, so their order doesn't matter. But in general we can't pull that kind of trick, even with simple events. For example, clearly if we're playing cards we'd be crazy to argue that because we know that

Pr(second card we draw is lower than an ace, given first card was an ace)

is pretty high (as there are so many such cards), then we can flip this around and bet heavily on getting an ace with our second card, because

Pr(second card is an ace, given first card was lower than ace)

must also also pretty high. The two events, 'first card is X' and 'second card is Y', clearly affect each other, and so aren't independent – so their order matters. Bayes gave us the means of flipping such 'conditional' probabilities around in all circumstances, and – crucially – it tells us that, to do it, we must also know the unconditional probabilities of the two events. So far, so simple; so what's the big problem? It comes when we start to use probabilities as measures of our degree of belief in something. Then the flipping process can lead to inferences that are not just silly, but dangerously misleading. Worried about the recurrent headaches you're getting, you go online and discover the disturbing fact that your symptoms are often associated with brain tumours, and that

Pr(getting headaches, given you have a brain tumour)

is around 50–60 per cent.[8] At this point, it's all too easy to just flip this around and conclude that

Pr(you have a brain tumour, given you're getting headaches)

is also around 50–60 per cent. Fortunately, however, having read this book you know that you can do that reliably only by using Bayes's Theorem, and that demands you take into account prior probabilities. In fact, wheeling out the full version of the theorem, we know that

Odds(brain tumour, given headaches) = LR × Odds(brain tumour)

where LR is the Likelihood Ratio, given by

Pr(headaches, given brain tumour)/Pr(headaches, given no brain tumour)

Now we see there's much less cause for concern, for two reasons. First and foremost, brain tumours are mercifully uncommon, being diagnosed in around one in several thousand people per year. So the prior probability we're one of them is also very low – making Odds(brain tumour) very low as well. But we could still have cause for concern if these low prior odds were boosted by a very high LR. We already have half the information needed to work that out: the 50–60 per cent figure for the chances of having headaches, if we have a brain tumour. Fortunately, however, that's only the top part of the LR: we also need the probability of getting headaches if we *don't* have a brain tumour. And as headaches are very common, this probability is also pretty high, and so the LR isn't. Upshot: low prior odds combined with unimpressive LR lead to low odds of having a brain tumour, given the evidence of the headaches. The lesson, then, is clear: whenever we want to know

Pr(our theory is right, given the evidence)

we need to be aware we could be making a huge mistake by thinking we can get this by simply flipping around the value of

Pr(the evidence we've observed, given our theory is right)

Yet, incredibly, that's the kind of trap researchers fall into whenever they use statistical significance to decide whether they've made an interesting discovery. In fact, it's worse than that – with consequences that have been devastating for the scientific enterprise. To see this, we need to do something that all too few researchers ever do, and get to grips with the probability at the core of the problem: the innocuously named 'p-value'. Happily, it's not difficult – though the implications are anything but happy for the scientific enterprise.

Introduced by Fisher in 1925 in his hugely influential *Statistical Methods for Research Workers*, the p-value appears to be a neat way of gauging the risk that a scientific result is really just a random blip. Clearly no scientist wants to make a big fuss about a fluke result, and Fisher suggested that the way to do it was to calculate the p-value, which he defined as the chances of getting findings at least as impressive as those obtained, *assuming* they were really just random flukes (see box opposite).

Fisher came up with the following rule, linking p-values to statistical significance: if the p-value calculated for a finding is below 5 per cent, then it can be deemed 'statistically significant'. All of which sounds fine, if a bit confusing. But there's a huge trap waiting for those who blithely accept all this. Fisher is saying a result is statistically significant if the chances of getting at least as

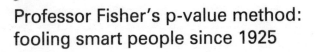

Professor Fisher's p-value method: fooling smart people since 1925

1. Calculate the p-value for the outcome of your study using formulas for:

   Pr(Getting outcome at least as impressive as that seen, assuming it's just a fluke)

2. If this probability is less that 5 per cent, call the result 'statistically significant'.

3. State outcome in your paper with its p-value, claim it as supporting your theory.

impressive a result, *assuming* it's really a fluke, are below 5 per cent. Yet why should anyone care about such a thing – and where does that figure of 5 per cent come from? Surely we should be looking at something far less convoluted, namely the chances that our results *really are* just a fluke – that is, calculating

Pr(Results obtained are a fluke, given the result we got)

and then checking *that* is below 5 per cent? Or how about forgetting all this stuff about fluke results anyway and just calculating

Pr(Results obtained reflect some genuine effect, given the result we got)

and seeing if that *exceeds* 95 per cent? Wouldn't *that* be a much clearer, intuitive and relevant definition of a 'significant' result? Indeed so – and notice how different it is from what Fisher is offering. It focuses on the actual results obtained, rather than the weirdly contrived 'at least as impressive results', and on whether these results reflect a genuine effect, rather than just one other rival explanation, namely that they're a fluke. But most worrying of all, Fisher's definition of a p-value has to be flipped around to get even close to what scientists should really be interested in. That is, his p-value is calculated on the *assumption* that fluke is the sole explanation of the results. As such, clearly we can't simply flip the p-value around and claim the very same number now represents the chances this assumption is incorrect. That's the classic flipping blunder, and is every bit as unreliable as assuming that since there's a high chance of having headaches given a brain tumour, there's exactly the same high chance of a brain tumour given headaches.

Yet that is exactly the blunder countless researchers have been making since Fisher first put forward his p-value test for 'significance' all those years ago. And the consequences have been filling up research journals ever since: bizarre results that wouldn't be taken seriously for a moment had they not passed his criterion for 'significance'.

So what possessed him to come up with so strange a definition? Put simply, his determination to avoid what Bayes tells us cannot be avoided: the introduction of prior knowledge and beliefs into the interpretation of scientific data. Professor Fisher was a brilliant mathematician, and knew full well the dangers of flipping around conditional probabilities with impunity. He also knew all about Bayes's Theorem, the problem of priors it raised, and how Bayes, Laplace and others had tried to tackle it. And he wanted nothing to do with it – least of all the idea of bringing subjective beliefs into the assessment of evidence. Fisher's loathing was visceral, although he often tried to disguise this using seemingly dispassionate technical reasons for rejecting Bayesian methods.[9] Having done so, Fisher had no choice but to concoct some non-Bayesian measure of use to researchers trying to make sense of their findings. The result was the p-value, whose notoriously contrived definition reflects its origins: as a doomed attempt to avoid the unavoidable. It's simply not possible to gauge the probability of a result being the result of fluke solely by using p-values. Fisher's use of the term 'significant' for results with low p-values looks suspiciously like a semantic ruse to dodge a mathematical fact. Certainly it created a risk of p-values being misinterpreted, and that's precisely what happened. Initially even Fisher himself fell into the trap of flipping low p-values around and interpreting them as implying a low chance of a result being a fluke. In fairness, within a few years of his textbook appearing, Fisher warned of the dangers of over-interpreting his concept of significance:

> *The test of significance only tells him [the practical investigator] what to ignore, namely all experiments in which significant results are not obtained ... Consequently, isolated significant results which he does not know how to reproduce are left in suspense pending further investigation.*[10]

In other words, Fisher was trying to finesse the role for p-values into one of simply weeding out junk not worth a second look. Even that was a shaky claim, however, and few working scientists seemed interested in it anyway. By the early 1950s, describing the 'complete revolution' Fisher's textbook was having on scientific research, one leading statistician expressed his concern that scientists were viewing 'significance' as the be-all and end-all of research.[11]

His concern was well merited. Despite repeated attempts, researchers have proved remarkably resistant to being disabused of their beliefs about statistical significance. There have been attempts to force the issue. In 1986, Professor Kenneth Rothman of the University of Massachusetts, editor of the prestigious *American Journal of Public Health*, told researchers that he would not accept results based solely on p-values. It had a dramatic effect: the number of papers relying solely on them plunged from over 60 per cent to 5 per cent. Yet two years later, when Rothman stepped down from the editorship, his ban on p-values was dropped – and researchers went back to their old ways. It's been a similar story in other fields, including epidemiology[12] and economics.[13]

Today, despite the occasional efforts of journals like *BASP*, little has changed. Academic societies have shown a remarkable reluctance to deal with an issue that 'retards the growth of scientific knowledge';[14] while several such institutions have looked at the issue, all have declined to take decisive action.[15] As a result, leading research journals continue to publish headline-grabbing 'statistically significant' claims which defy credulity or attempts at replication. Meanwhile new recruits to the scientific enterprise are taught significance testing – often with textbooks using flawed definitions and no warning of the real meaning of it all. Research has shown that many students who think they know what p-values are actually don't.[16]

The result has been decades of wasted time, money and effort by researchers – and declining confidence in scientific claims among the rest of us.

↑UPSHOT

To find out whether an experimental finding is worth taking seriously, scientists routinely apply so-called significance tests – despite repeated warnings that these methods are fundamentally flawed and dangerously misleading. The result has been a plethora of unreliable 'breakthroughs' – and growing concern about the reliability of scientific claims among both researchers and the public.

Dodging the Amazing Baloney Machine

Of all the sciences, physics is generally regarded as the hardest – and not just in the sense of being intellectually demanding. Its theories have a reputation for being rock solid, based on deep insights into the design of the universe. The extent to which that reputation is deserved is moot; what isn't is the extent to which physicists have made triumphant use of 'Big Data'. While researchers in the 'softer' sciences like psychology often have to make do with analysing questionnaires from a few dozen college students, physicists like to test their cosmic theories using data-points counted in billions and trillions. And no one does this better than experimental particle physicists. Their goal is to uncover the secrets of the basic building blocks and forces of the cosmos, and their weapons of choice are machines like the 27-kilometre-long Large Hadron Collider (LHC) at CERN, the European nuclear research centre in Geneva, Switzerland. Their modus operandi involves smashing together hundreds of billions of subatomic particles a second for hours on end, and scouring the debris for telltale signs of their quarry. The reason they need so much data is that what they're looking for is often incredibly rare. But over the decades they have become masters at finding specks of scientific gold in mountains of random dross – and have the Nobel Prizes to prove it. In December 2011, the CERN team made headlines by discovering the long-sought Higgs particle, a key part of theories unifying all the forces and particles of nature. Calculations suggested it would reveal its fleeting existence in perhaps one in a billion collisions. Among those would be countless random events faking the presence of the Higgs. Even so, after checking the outcome of over 100 million million collisions, the team were able to announce that they'd found the particle predicted by theorists over fifty years earlier.

The discovery of the Higgs was a hard-won triumph. It was built on sometimes bitter experience of the tricks randomness can play on the unwary – and of the inadequacy of the methods conventionally used by scientists to deal with them. Had the CERN team followed the traditions of researchers in other fields and declared their discovery using the standard methods of significance testing, the 2011 announcement would have been greeted with eye-rolling scepticism. That's because it would have been just the latest of claims to have found the Higgs dating back to the mid-1980s. Fortunately – and in stark contrast to other areas of science – researchers in particle physics have long insisted on far more stringent standards of evidence before going public with supposed 'discoveries'. Certainly no one at CERN was keen to repeat the debacle of 1984, when the laboratory went public with claims to have found another key component of the cosmos that proved to be a random fluke. Analysis of accelerator data had pointed to the existence of the so-called top quark, with a mass around forty times that of the proton.[1] The team's confidence seemed justified given that the evidence comfortably passed the standard adopted in other areas of science for declaring a result 'statistically significant' – seemingly implying it was unlikely to be random noise. Yet as more evidence emerged, the discovery proved to be just that; other discoveries claimed by CERN and a rival laboratory that same year went the same way.[2] The debacle underlined long-standing suspicions among particle physicists about the reliability of statistical significance as a measure of evidence. A decade later, a rival team in the USA was again claiming evidence for the existence of the top quark, but this time on the basis of a far higher standard of evidence. The claim has since been confirmed many times – as has the mistaken nature of the CERN 'discovery': the real top quark has a mass over four times that claimed in 1984.

While particle physicists are best known for their use of giant machines like the LHC, they owe much of their success to their scepticism about the Amazing Baloney Machine in the guise of 'significance testing', whose output is routinely relied on by scientists in other fields. For decades particle physicists have witnessed the unnervingly common disappearing act performed by claims passing the test laid down by Ronald Fisher in the mid-1920s: that results with p-values of less than 5 per cent can be deemed 'significant'. As we saw in the previous chapter, researchers routinely make the mistake of assuming that this means the chances of their result being a fluke are also below 5 per cent. Fed with this assumption, the Amazing Baloney Machine then turns meaningless flukes into 'discoveries' whose real nature becomes apparent only if anyone tries to confirm them. Particle physicists have tried to see off the machine's worst excesses by feeding it with more impressive levels of significance, usually packaged up in so-called sigma units, a neater and more intuitive way of expressing the same thing as p-values.[3] Fisher's $p = 5$ per cent standard for declaring a result 'significant' now became a '2-sigma result', with higher sigma values indicating higher levels of significance. But over the years, physicists noticed that even 3- and 4-sigma findings – corresponding to much more 'significant' p-values of 0.3 per cent and 0.006 per cent – also had a habit of fading out in the face of more data. By the mid-1990s, 5-sigma had become the minimum significance level accepted by the leading physics journal for any claim to have made a discovery. And by the standards of conventional science, this is a breathtaking demand, corresponding to a p-value almost 80,000 times more 'significant' than Fisher's 5 per cent level commonly used. Yet particle physicists are wary of what emerges from the Amazing Baloney Machine if it's fed with anything less. Within the community, this scepticism is captured in a rule of thumb: 'Half of all 3-sigma results are wrong'.[4] This is an intriguing observation that hints at the source of the problem in trusting the Machine. If significance tests meant what so many researchers think they do, then the underlying theory of sigma values would mean that 3-sigma results prove to be meaningless fluke on average only once in every 370 cases. Yet according to that rule of thumb, the true rate is closer to one in two. Of course, random flukes aren't the sole reason experiments prove unreliable. Simple mistakes are quite capable of undermining claims. In 2011 reports emerged of particles called neutrinos travelling faster than light. The data surpassed the 5-sigma 'discovery' level, so the finding seemed unlikely to be a fluke – and indeed it wasn't: it proved to be the result of faulty equipment. Nevertheless, the 100-fold mismatch between what researchers *think* the Amazing Baloney Machine is saying and what they actually *get* suggests there's something seriously wrong in their understanding of the machine. And, as we saw in the previous chapter, there is: they are expecting the machine to perform miracles – namely, to take raw data, calculate probabilities like

Pr(Observing at least as many hints of Higgs, assuming they're flukes)

and then flip it, in the hope that the very same number is the answer to the key question

Pr(The hints are just flukes, given how many we've seen).

Bayes's Theorem tells us such flips are a very risky manoeuvre unless we've got other information – in particular, the prior probabilities for what we're investigating. Given those, it can give the answer to the key question that the Amazing Baloney Machine seems to provide, but just can't. But Bayes can also tell us just how big a mistake we may be making by trusting the Machine, and the results are telling.

Take the most common p-value mistake: of flipping a 'statistically significant' 2-sigma result (equivalent to Fisher's p-value standard of 5 per cent) and assuming it means the chances of our result being a fluke are also just 5 per cent. Bayes tells us we can only do this if we have some prior insight into the risk of our result being a fluke. As ever, it also confirms the common-sense notion that the less compelling the evidence, the more convinced we have to be beforehand that our results aren't a fluke. But cranking through the maths,[5] a shocking fact emerges. It turns out we're only justified in interpreting the classic 'p less than 5 per cent' result as implying a less than 5 per cent risk of a fluke if we're *already* 90 per cent sure that fluke can't be the explanation. In other words, the evidence of the prototypical 'significant' result is so feeble it adds virtually nothing to our existing level of belief.

In reality it's not just physicists who have become sceptical about claims based on p-values close to the traditional 5 per cent cut-off for statistical significance. Bitter experience has taught researchers in many fields that Fisher's criterion for significance is just not good enough. This has led many to tackle the problem in the same way as physicists, by demanding more impressive evidence – p less than 0.1 per cent, or at least 3-sigma – before taking new results seriously. Bayes confirms that this helps – but not much. Despite appearing to be 50 times more impressive than Fisher's standard, even this level of evidence still demands we *already* think there's no more than a 30 per cent risk that fluke can't be the explanation before taking it seriously – in the sense of regarding the fresh evidence as pushing the risk of fluke below 5 per cent, as the p-value seems to imply. And the fact is that researchers in most disciplines rarely get anything like this level of evidence.

The good news is that Bayes can do more than merely blow holes in the Amazing Baloney Machine. It can give us some rules of thumb to make sense of what comes out of the machine. And to be fair, Fisher's machine does at least try to give those who feed it what they want. In particular, the 5 per cent cut-off for significance specified by its illustrious designer has proved popular with researchers. So let's take that 5 per cent level, and build a Bayesian version of the machine around it so that it means what it seems to: that the evidence implies just a 5 per cent risk of a result being a fluke. Of course, the Bayesian machine will need to be fed with data, but it will also need our prior level of belief – the key ingredient absent from Fisher's machine.

What we'll call the Bayesian Inference Engine is, like Fisher's machine, a formula,[6] and it leads to the following rules of thumb. In each case, it gives us a rough indication of the *minimum* level of prior belief that the result isn't a fluke needed for us to take seriously the various levels of evidence. Here 'take seriously' means the evidence meets the time-honoured standard of no more than a 5 per cent risk of fluke. The table also includes subject areas where such levels of evidence are often stated as being at least suggestive, if not 'significant' or even compelling.

Level of evidence (p-value)	Typical areas where such levels of evidence turn up	How convinced you *already* need to be to find this level of evidence impressive
10 per cent	Economics, sociology, 'controversial' health/environment/risk issues	95 per cent Only those *already* convinced will be impressed
5 per cent	Almost ubiquitous; especially prevalent in behavioural, social and medical sciences	90 per cent Impressive only if you think it very unlikely to be a fluke
1 per cent	Medical sciences, genetics, environmental sciences	75 per cent Impressive only if you're fairly sure result can't be a fluke
0.3 per cent	Lab studies in 'hard' sciences; preliminary ('3-sigma') claims in particle physics	50 per cent Capable of impressing open-minded agnostics
0.1 per cent	Genetics, epidemiological studies	30 per cent Impressive to all but moderate to severe sceptics
0.00006 per cent	Discovery claims in high-energy and particle physics ('5-sigma')	0.1 per cent Very likely to impress all but your rivals

The most striking thing about the Bayesian Inference Engine's output is just how thin most supposedly 'significant' evidence proves to be. As the table on the previous page shows, such evidence typically demands we're *already* pretty sure the findings aren't a fluke before we're justified in taking them seriously. And that, remember, means 'seriously' only in the sense of believing there's a 5 per cent chance that they're flukes. If we set the bar higher – demanding just a 1 in 100 chance of being fooled by fluke, say – the level of evidence needed soars.

Perhaps the most telling result is that it takes a pretty stringent (and far from common) p-value of 0.3 per cent before even an open-minded sceptic can be convinced that fluke can been ruled out. Anyone harbouring greater scepticism must demand even more impressive evidence before feeling confident about setting fluke aside as an explanation.

As ever, one shouldn't forget that fluke alone isn't the only reason that findings can prove misleading. Indeed, research claiming to have staggeringly low p-values – and thus staggeringly high sigma and significance levels – has a reputation for being pretty strong evidence for ESP: not Extra-Sensory Perception (although, ironically, it often does appear in that context), but 'Error Some Place'. Science is a human endeavour, and so will always reflect human foibles. The Bayesian Inference Engine cannot fix them all, but it can save us from the folly of following researchers proclaiming as 'significant' results that are, on any reasonable definition of the term, anything but.

↑UPSHOT

Many headline-grabbing 'breakthroughs' in science are based on 'statistically significant' findings. Bayes's Theorem leads to simple rules of thumb for making sense of such claims – and all too many prove to be based on evidence so weak it should impress only those who are already 'true believers'.

Making use of what you already know

If you want to make sense of new findings, the Bayesian Inference Engine gives straight answers to straight questions – which is more than can be said for the Amazing Baloney Machine and its p-values. So why is anyone still using what one eminent researcher memorably described as 'surely the most bone-headedly misguided procedure ever institutionalised in the rote training of science students'?[1] One reason immediately becomes clear to anyone who flicks through textbooks on Bayesian methods. Most are packed with heavy-duty maths, with seemingly little interest in dealing with such trivia as 'Are my findings just a fluke, or what?' That's because while Bayes gives straightforward answers to such questions, getting to those answers can involve maths so complicated it can only be done with the aid of computers.[2] For many years, this was a major barrier for those wanting to ditch the Baloney Machine, but it's now been overcome, with standard software packages available to do the heavy lifting.

Even now, however, many would-be users of Bayes's Theorem are deterred by the centuries-old 'Problem of Priors'. How do we reach an initial level of belief, even before we've seen the data – and doesn't it allow subjectivity to seep into the scientific enterprise? At least, that's how it's usually talked about. But just how big a 'problem' is it? Is not an ability to take account of what we already know actually an advantage? For the fact is that, after decades of research across so many fields, we actually have some pretty good insights into many things, and Bayesian methods allow us to make use of this, and set new results in context.

The trouble is that all those past insights sometimes strip the headline-grabbing gloss off claims for 'major breakthroughs' and 'miracle cures' – and no one likes a party-pooper. Just ask Allen Roses, a senior executive at the pharmaceuticals company GlaxoSmithKline (GSK), who in December 2003 found himself making front-page news after admitting that, despite spending billions on the search for new therapies, the vast majority of drugs don't work on most people.[3] As the reporter who broke the story pointed out, this was hardly news to those involved in the search for new treatments. They've long known that despite all the hype about the wonders of modern medicine, 'miracle cures' are few and far between, and claims to the contrary needs to be viewed with suspicion.

Despite this, in deciding whether or not to approve some new therapy, regulators still put their trust in the techniques of significance testing – which offer no means of taking explicit account of past experience. In contrast, the Bayesian Inference Engine cheerfully accepts both the data from studies and insights from past research before giving an answer. And if the claim being made flies in the face of past experience, it can sound the alarm about potential disappointment ahead. In September 1992 medical researchers in Scotland made headlines with the results of a study of a drug called anistreplase. As a 'clot buster', this drug belonged to a family already transforming the survival prospects of patients with heart attacks, who were given the drug as soon as they arrived at hospital. Given the benefit of speed, however, it seemed entirely plausible that the drug would save even more lives if administered by a doctor even before the patient got to hospital. The Grampian Region Early Anistreplase Trial (GREAT) was set up to find out, and its findings were dramatic: death rates among heart attack victims given the drug before reaching hospital were half those of patients who got the drug in hospital. Given how common heart attacks are, this seemed a major breakthrough. But the response from experts was muted. They pointed out that while some benefit made perfect sense, so large an improvement flew in the face of previous experience. Even so, by the usual standards of assessing evidence, the GREAT findings passed muster: they came from respected researchers, and were statistically significant, with a p-value of 4 per cent – just within the time-honoured 5 per cent limit.

In the years that followed, other teams set about replicating the breakthrough, and in 2000 a review of all the evidence was published, based on over 6,000 patients – 20 times the size of the GREAT study. The good news was that the technique did indeed seem to offer some benefit; the less good news was that overall it seemed to produce only a 17 per cent reduction in risk of death – still worthwhile, but considerably less effective than the original study had suggested. In short, the GREAT study seemed to be another case of the fading breakthrough – and as ever there was no shortage of potential explanations, not least the relatively small size of the initial study. But one explanation stood out – one which predicted not only that the original findings would fade, but also by how much.

Shortly after the study had been published, two British medical statisticians, Stuart Pocock and David Spiegelhalter, wrote a short letter to the *BMJ* arguing that the halving of the death rate needed to be *put in context*.[4] But rather than resorting to the usual vague generalisations, they set about doing it in quantitative detail, using Bayes's Theorem.

Put simply, they argued that the new study should not be seen as a stand-alone result, its reliability judged solely by significance tests. Instead, they pointed out that it constituted fresh evidential weight that could be combined with prior insights about clot-busting drugs and the likely impact on death rates. Pocock and Spiegelhalter captured this prior knowledge in a so-called 'credible prior interval' – that is, a range of values within which the real risk of death was likely to lie, in the light of current knowledge (see box opposite).

When combined with the new research findings, they found that the real life-saving effect of the clot-buster was more likely

How Bayes showed GREAT wasn't so great after all

To provide a simple summary of their findings, researchers often use so-called confidence intervals (CIs), which give a 'headline figure' and a plus-or-minus range, reflecting the effect of chance. So, for the GREAT clot-busting study, the team summed up the findings with a 95 per cent CI for the relative risk of death for those who got the treatment compared to those who didn't of 0.47 (0.23 to 0.97). As no relative benefit would give a value of 1.0, it seems the treatment produced a 53 per cent (= 100– 47 per cent) cut in the risk of death, with a 95 per cent chance of the benefit being as much as 77 per cent, or as little as 3 per cent. The 95 per cent standard is used by analogy with the 5 per cent p-value standard. Yet as with p-values, the correct interpretation of a 95 per cent CI is both technical and not the answer to the question we want answered – and it takes Bayes's Theorem to make things clearer and more relevant. Put simply, standard CIs give only 95 per cent 'confidence' of including the true value if we presume utter prior ignorance of what the true value might be, and also assume only chance can undermine the finding – two pretty big caveats.[5]

Despite still being somewhat misleading, 95 per cent CIs are certainly better than p-values because they do contain more information. If the range excludes values corresponding to no effect – which in the GREAT case mean a value of 1.0 – then the result is said to be 'statistically significant'. As we've seen, that doesn't mean much. Far more significant, however, is the *width* of the CI – that is, the difference between its upper and lower limits. Small samples are more vulnerable to chance effects, and reveal themselves as wide CIs. In Bayesian terms, these in turn imply low evidential weight – and the GREAT results were a case

in point. When Pocock and Spiegelhalter used Bayes to combine
the study's feeble weight of evidence with the results from two
much larger studies pointing to less dramatic effects, its headline
figure of a 53 per cent cut in death risk shrivelled to 25 per cent –
which, years later, turned out to be more realistic.

to be around 25 per cent – still worthwhile, but much less than
suggested by the study alone. The authors struggled to get their
calculations published, but when the review of the evidence was
published seven years later, pointing to a 17 per cent reduction,
their Bayesian prediction was vindicated.[6] It was an impressive
demonstration of the importance of taking into account past experience and *plausibility* when making sense of new findings. And
crucially, having published their prediction of potential disappointment years before the results were in, Pocock and Spiegelhalter
could not be accused of having benefited from hindsight.

Yet at the same time, they had highlighted some important
questions about the use of Bayes in matters of life-or-death importance. Didn't their calculation prove that Bayes really does allow
everyone to reach their own hand-picked conclusions? Suppose,
for instance, that they had been rivals of the original researchers,
determined to kill off their clot-busting research. What was to stop
them carefully selecting prior evidence and running it through the
Bayesian machinery until the trial results looked stupid? If they had
been been cheerleaders for the treatment, or on the payroll of the
clot-buster makers, they could just have easily had tilted the findings the other way.

Such criticism would carry more weight were it not for the fact
that researchers have always dismissed or embraced new findings
on the basis of their personal insights, or prejudices – or venal-
ity. Lunch breaks at research institutes are regularly enlivened by
arguments over new, headline-grabbing results, with liberal use of
phrases like 'Well, I still don't believe it' and 'You've got to admit, it

does make some sense'. The use of significance tests does nothing to stamp out such blatantly subjective practices. That's because every researcher knows from experience that, no matter how impressive the p-value, if a result doesn't 'smell right', scepticism is often still merited. What significance testing actually does prevent, however, is any hope of putting this on a transparent and *quantitative* basis. Instead, sceptics and true believers alike can get away with vague, hand-waving justifications. And not just over lunch, either: to read the 'discussion' sections of papers in prestigious journals is to be exposed to untrammelled subjectivity dressed up as expert insight. Pocock and Spiegelhalter's principal achievement in that short letter to the *BMJ* was to show it doesn't have to be this way. Bayes's Theorem puts the process of putting new results in context on a solid, mathematical foundation. Of course, it's entirely possible to simply pick and choose what prior evidence to combine with the new findings. The crucial difference is that Bayes's Theorem compels sceptics and true believers alike to state *explicitly* what prior evidence they are bringing to their assessment.

The idea of sullying pristine findings with possibly faulty prior results may still seem risky, but the Bayesian Inference Engine has this covered. Its underlying mechanics ensure that as the data accumulate those prior beliefs become ever less important. Unless it's fed with some pretty weird prior beliefs, it will lead both sceptics and true believers to the same conclusion – which no amount of argument over lunch may ever achieve.

↑UPSHOT

Assessing the plausibility of new findings means putting them in the context of what we already know. All too often, the result is barely more scientific than 'That sounds reasonable'. Bayes's Theorem gives us a consistent, transparent and quantitative means of gauging the plausibility of new findings.

I'm sorry, professor, I just don't buy it

The scientific method has many astounding accomplishments to its credit. Orbiting observatories have shown that the universe began in a Big Bang around fourteen billion years ago. Clinical trials have given us effective treatments for a host of killer diseases. And brain scans of men watching porn show it makes their brains shrink.[1]

Hardly a week goes by without the media reporting some more or less bizarre claim based on research published by real scientists in serious journals. Such is their prevalence – and apparent credibility – that in 2007 the UK National Health Service set up an online site, *Behind the Headlines*, where experts analyse the claims and put them in context. Unusually, the site does not make the a priori assumption that all journalists are unreliable merchants of hype, nor that all researchers are brilliant seekers after truth. Instead, they stick to explaining what has been claimed, and the extent to which it can be justified. And with all too many studies, the answer is: hardly at all. From studies pointing to the miraculous or murderous effect of eating eggs to research backing the idea of 'gaydar' that allows people to tell whether others are homosexual,[2] many of the studies make the headlines precisely because they're about questions never before addressed. And virtually all of them reach their conclusions via the standard ritual of feeding raw data into the Amazing Baloney Machine.

But how could Bayes help in such cases? After all, to work at all, the Bayesian Inference Engine needs not only raw data, but also prior insights – and where are they to come from when no one's done anything similar in the past? It's a challenge made all the more difficult by the fact that such out-of-the-blue studies are often small. As a result, they don't pack a lot of evidential weight, and what there is could be swamped by a badly chosen prior.

We're facing the centuries-old Problem of Priors again, and this time it seems especially serious. One way out is to wave the white flag and turn our Engine into the Baloney Machine – by feeding it with a so-called 'uninformative' or 'vague' prior. That amounts to assuming that all outcomes – no matter how silly – are equally likely. A less abject response is to accept that we don't have obviously relevant prior evidence to use – and instead seek insight from more broad-brush if less precise sources, or, as they're often called, 'experts'. This involves a process known as prior elicitation, which at its simplest involves getting experts to guesstimate plausible ranges within which they expect the true result to lie. For example, they could be asked to state a 'most likely' effect size, along with an estimate of the highest plausible level. These can then be combined to produce an overall 'expert prior distribution', which can be fed into the Engine to set the study outcome in context. It's a process that clearly has its dangers, however. Experts can and do produce wildly inaccurate guesstimates,[3] and these can seriously affect the interpretation of small studies. And in any case, what if we don't agree with the experts, or they're later shown to be wrong? How do we unpick their influence on the interpretation of the study?

Fortunately, there's a button tucked away on the gleaming exterior of the Bayesian Inference Engine that even many veteran users have overlooked. It allows us to avoid feeding it with either hopelessly vague or misleading 'expert' prior insights – and reach our own, personalised view of the evidence. In essence, pressing the button puts the engine into reverse. Recall how it usually operates: it starts with prior insights, combines them with the raw data we've obtained and then tells us whether the evidence is now compelling, in the light of what we already know. But the engine works equally well in reverse. That is, it will cheerfully *start* with what we'd regard as a compelling conclusion, and work backwards to reveal the level of prior belief we need for the data to justify such a conclusion. Thus, instead of insisting – somewhat ludicrously – that 'no one knows anything' or that – rather snootily – only 'experts' should set priors, pressing the button on the engine allows every one of us to make sense of the data in our own terms. The engine tells us what prior belief we must hold for the data to lead to a compelling conclusion. All we have to decide is: do we find that level of prior belief reasonable? We may find it too outlandish, in which case we're entirely justified in regarding the new findings as unconvincing. If, on the other hand, we have no problem squaring it with our own belief, we're equally justified in claiming the new research has made its case. The whole process is transparent, democratic and quantitative – and for many types of study, involves simply feeding two numbers into an online calculator.[4]

Even running in reverse, the engine retains all of its power – including its ability to reveal the true strength of the evidence. Take the case of the GREAT heart attack study in the previous chapter, with its impressive claim of a 50 per cent reduction in death risk if the treatment is given quickly. The engine makes short work of the 'statistically significant' headline finding of a halving of mortality. Put into reverse, it reveals that in order to regard that result as convincing, one would *already* need to be convinced that early treatment would produce *at least* a 90 per cent cut in mortality. That's because the weight of evidence from the GREAT trial is so feeble – and thus its findings don't add much to whatever prior knowledge one might have. Indeed, the GREAT study fails to make its case even in its own terms: its evidential weight is so low that its 50 per cent figure is convincing only if there's already evidence for a far more impressive outcome. Does that mean the study was a waste of time and money, and half of the patients were put in harm's way for no reason? Absolutely not: the whole point of research is to push back the boundaries of knowledge by accumulating evidence. The GREAT study was a crucial part of that process, and the Bayesian Inference Engine makes the most – and the most sense – of what such studies are telling us. Sure enough, as more studies were conducted into this approach to saving lives, more evidence accumulated – and the engine showed it became ever more convincing. When the evidence was reviewed, with the resulting headline figure of a 17 per cent cut in death risk, based on 20 times as many patients as the GREAT study, that figure carried far more evidential weight, and thus a much narrower 95 per cent confidence interval. When it is put into the engine set into reverse, we find that the credibility of this new finding no longer demands that we *already* believe a 90 per cent cut in deaths is possible. Now we're only required to think a 28 per cent cut is plausible in order to take the new evidence seriously – which is far less demanding. The engine is showing that the new data are strong enough to do most of the heavy lifting now, and don't need so much help from prior knowledge.

The engine can help make sense even of the most perplexing form of evidence: that which comes 'out of the blue' from research into completely new questions. Such studies leave even experts groping for anything meaningful – let alone quantitative – to say. For example, in 2012 a team at the University of Miami went public with the claim that people who daily consumed diet soft drinks faced a hefty – and statistically significant – 43 per cent increased risk of vascular events such as stroke.[5] Given the popularity of 'low cal' drinks and the fact that the study involved thousands of people, the claim made headlines even before it was officially published. Yet the researchers themselves worried about their findings being pushed too far. They stressed that despite the size of the overall study, the headline-grabbing figure was based on a subset of less than 10 per cent of the participants. The researchers duly called for much bigger studies of this potentially important finding. What neither they nor anyone else did, however, was to do more than feed the raw data into the Amazing Baloney Machine. If they had, they would have realised just how feeble the evidence actually was. Feeding the data into the Bayesian Inference Engine after we've put it into reverse, it tells us that the headline-grabbing figure of 43 per cent is credible only if we're already convinced that the true figure is at least 60 per cent. But given that this is the first ever published study to make any such claim, where should such a belief come from? After all, not even the study itself is claiming so dramatic a risk figure. In other words, the study lacks so much evidential weight that – like the GREAT study – it's not even credible *in its own terms*. The engine is warning us that what we have here is statistically significant evidence in its most feeble form, in that we must already believe in an effect size more impressive than the study itself found in order to regard it as credible. Sure, the study has contributed some weight of evidence, and that's potentially a useful contribution to science. But it's also far more tentative than that hefty risk figure and its 'statistically significant' sticker implies.

As the time-honoured phrase goes, more research is needed. In the meantime, we should ignore the media coverage, leave the scientists to find out more – and perhaps instead ponder the following. Since their invention in the 1920s, significance testing and p-values have been confusing students, fooling researchers and misleading the rest of us into seeing 'significance' in results that are anything but significant. Ironically, having been invented as a gentle way of weeding out obvious flukes, they have been transformed into the Amazing Baloney Machine, which claims to reveal what to take seriously, but in fact cannot. Whether it's the results from the latest investigation of a widely studied medical treatment, or an out-of-the-blue claim about something no one's studied before, it's all the same to the machine. It just takes the data in, ignores everything else – and gives its pronouncement: either 'gold dust' or 'garbage'.

Such an approach is inimical to scientific progress. At every level, from the discovery of the expansion of the universe to the identification of the genetic role of DNA to the demonstration that protons contain quarks, science has advanced by the accumulation of evidence, not by simple true/false dichotomies. Scientists capture reality in subtle shades, not black and white. And the way to do this is by combining different hues of evidence using Bayesian methods.

Even now, with so much evidence having built up for the failings of the Amazing Baloney Machine, such statements are apt to provoke paroxysms of outrage. But those determined to keep faith with the machine are setting their face against the outcome of a research programme that began just as the machine was being built. During the 1920s, a number of mathematicians – notably Émile Borel in France, Frank Ramsey in England and Bruno de Finetti in Italy – began pondering the question of how hard evidence is turned into the fuzzy thing called belief. Their work uncovered the laws that any rational and reliable approach must follow. And they proved to be the laws of probability – with Bayes's Theorem playing the key role of updating belief in the light of evidence. Largely ignored for decades, this intriguing link was explored by others seeking to put it on a rigorous basis.[6] In the last few years, the fundamental roots of the connection between inference and Bayes's Theorem have been found, and the connection turns out to be not merely plausible, but actually ineluctable.[7]

There is, in short, no longer any excuse for keeping faith with the Amazing Baloney Machine. It needs to be wheeled off to the scrapyard before it does any more damage to the scientific enter-prise. But there are some parts of it that could be salvaged. There's no doubt it has one very appealing feature – one which doubtless explains its enduring popularity: the machine may have given mis-leading guidance about the 'significance' of new evidence – but at least it was clear guidance. The good news is that we can still get this from the Bayesian Inference Engine.

What we must consign to the breaker's yard of science, however, is the idea of a simple pass/fail test. It's time we all – from con-sumers of scientific evidence to its creators – bought into a more nuanced approach to evidence.

↑UPSHOT

The Bayesian Inference Engine allows us to put new evidence into context, allowing us to update what we know. But it can also help us make sense of out-of-the-blue research in fields where next to nothing is known – and to spot when evidence is so feeble it's not telling us anything much at all.

The Amazing Curve for Everything

When TV producers want someone to appear intelligent, they'll ensure there are some shelves full of books in the background. When they want the person to look like a genius, they'll replace the shelves with a whiteboard covered with mathematics. They've long recognised how the mere appearance of a few equations dispels doubt and conveys authority. Mathematicians themselves aren't unaware of the power their strange language can exert on non-speakers. According to legend, in 1774 the brilliant Swiss mathematician Leonhard Euler won a public debate on the existence of God by scribbling a meaningless formula on a blackboard, declaring it to be proof that God existed, and demanding a response. Utterly baffled, his innumerate adversary fled the room. While the story is apocryphal,[1] it speaks to a larger truth: that one of the most effective ways of quelling dissent is to declare: 'There's some algebra for that.'

This may help explain why in the late 1990s senior managers at some of the world's biggest companies became infatuated with the following serving of Greek alphabet soup:

$$f(x, \mu, \sigma) = \frac{1}{\sigma\sqrt{2\pi}} \exp\left(-\frac{(x-\mu)^2}{2\sigma^2}\right)$$

Getting on the wrong side of this little lot can cost you your job

For over a decade, employees at the likes of Microsoft, General Electric and Conoco could and did find themselves fired for ending up on the wrong side of this formula – or, more precisely, the curve it describes, shown below:

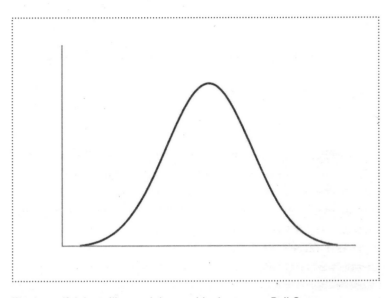

The beautiful, beguiling and thoroughly dangerous Bell Curve

It's the famous Bell Curve, and for a while human resources departments became convinced it captured with mathematical precision the performance of employees, measured according to pretty much any metric you fancied: sales, profits, 'effectiveness', whatever. The curve captured graphically the supposed truths embodied in the formula. First, that most employees are near-average performers, and lie near that central 'hump', with half of the staff above average, and the other half below. Secondly, a small proportion of employees are real stars, with exceptional performance putting them in the right-hand 'tail' of the Bell Curve. And thirdly, there was a matching proportion of slackers, losers and deadbeats, all huddling in that left-hand tail. These could be identified, hauled in for a pep talk, or fired. But how to do it? Simple: rate the performance of the staff on a scale of 1 to 5, making sure the proportions getting each rating follow the dictates of the Bell Curve. So most should get the average score of 3, while somewhat fewer should be rated 2 or 4. With those dealt with, management could then focus on the 'outliers'. Those in the right-hand 'tail' rated 5 would get the bonuses, while their counterparts in the left-hand side would get the boot.

Unsurprisingly, this bizarre routine triggered considerable resentment among employees – and suspicion. Many sensed there was something not quite right about what became known as 'Rank and Yank'. Some of those finding themselves in the wrong tail of the Bell Curve decided to take their employees to court. Yet few felt confident about taking on the formula itself. Astonishingly, it took over a decade before its mathematically induced spell was broken. Yes, the formula is mathematically correct, and yes, there's no doubt the Bell Curve reflects many human traits, such as height and IQ. But no one thought to check whether 'performance' was one of them. When they did, the results confirmed what many people had suspected: that the spread is anything but symmetric.[2] Instead, it's usually a case of few top performers, plus everyone else. The idea that there *must* be equal proportions of stars and slackers in every department turns out to be – unsurprisingly – beyond stupid, and a serious threat to corporate well-being. By forcing ratings to conform to the dictates of the Bell Curve, managers found themselves compelled to reprimand, say, 10 per cent of employees simply because 80 per cent had to be found at or near average – leaving 20 per cent in the two 'tails'. In the end, the lack of evidence that it did anything but wreck morale has led many former advocates to abandon Bell Curve appraisals. Microsoft and many others have moved on, but many persist. Some may well have good cause, but chances are they remain stuck in one of the deepest traps in dealing with uncertainty: a belief that pretty much everything is Normal.

That might seem a perfectly reasonable belief, but the capitalisation here is crucial. For like so many terms in the theory of chance and uncertainty, Normal has a very specific meaning that almost invites abuse. It seems to imply ordinary, standard or natural, but in this case it means compliance with the dictates of the Bell Curve – or, as mathematicians call it, the Normal Distribution, whose formula was given earlier. Indeed, the term is doubly inappropriate, for not only does the Normal Distribution often fail to describe 'normal' phenomena, but the formula behind it is the outcome of one of the most exceptional mathematical discoveries ever made.

Its roots stretch back to the very dawn of probability theory. During the seventeenth century, the field's pioneers – among them Pascal, Fermat and Bernoulli – had found ways of working out the chances of different combinations of events, such as getting three 6s during ten throws of a die. The answers emerged from formulas which included both the chances of the individual event happening in a single attempt, and the number of ways ('permutations') in which that one-off event could appear during the throws. For example, three 6s could appear on the trot, or at random intervals. Something intriguing emerged when the results were plotted on paper, however: as the total number of attempts increased, the chances of getting a specific number of successes seemed to follow a distinctive curve.

It emerged even from the simplest manifestations of chance, such as a coin-toss. Given that the chances of getting heads on any one toss are 50:50, we'd expect the most likely number of heads to be half the total number of tosses. Sure enough, plotting out the results from the formula in the case of ten tosses produces a peak in probability at five – the average number of heads we could expect to see. The formulas also gave the chances of getting other numbers of heads from ten tosses – and these showed up as steep slopes to either side of the central peak, reflecting their lower probability of occurring. And at either extreme were the rarest events of all: either no heads, or all heads, from ten tosses.

Calculating these graphs by hand is not for the faint hearted. Even a master mathematician like Jacob Bernoulli struggled to deal with anything other than small numbers of trials.[3] Yet without being able to perform such calculations, it was hard to find out much more about these curves. What was needed was some kind of short

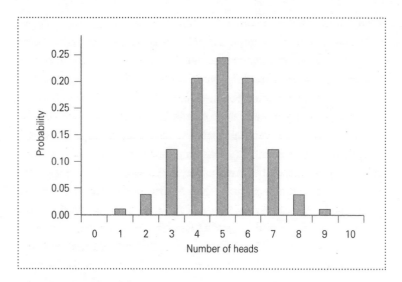

The Bell Curve rises: the chances of getting different numbers of heads from ten coin-tosses

cut, and in 1733 the breakthrough came in the form of a formula that was not just easier to use, but even became more reliable as the number of trials increased. It had been devised by Abraham de Moivre (1667–1754), an émigré French maths teacher and consultant living in London and one of the most brilliant mathematicians of his day. De Moivre's skills in the theory of chance were such that even that imperious genius Isaac Newton allegedly deferred to him in such matters. Ironically, he was also somewhat unlucky, missing out on being credited for several discoveries – including his neat formula for probabilities. Instead, it acquired various monikers, including the Gaussian curve, after the great German mathematician Carl Gauss (1777–1855), who had found the formula via a completely different route. At the time, Gauss was wrestling with one of the central problems of experimental science: extracting insights from data subject to error. He showed that by making three reasonable assumptions about how errors affect observations, he could calculate the chances of the true value lying in a specific range. His formula was essentially the same as that found by de

Moivre, and it's the one found at the start of this chapter. When plotted out, it also gives the Bell Curve. De Moivre had already shown that the central peak coincided with the most likely outcome of a given number of random events such as coin-tosses. This was handy for gamblers wanting to work out the chances of a bet paying off. But Gauss had shown that the peak also marks the average of a set of measurements, each subject to random error. And that made it hugely useful to scientists trying to gauge the likely range within which the true value of some quantity might lie.

The formula's first public outing made Gauss internationally famous at the age of just 24. On 1 January 1801, an Italian astronomer created a sensation by claiming to have found a new planet in the solar system, orbiting between Mars and Jupiter. Unfortunately, before anyone could confirm the discovery, the object was lost in the glare of the sun. Without knowing its orbit, there was a risk the planet might not be found again for years. Gauss applied his formula to extract the maximum value out of the existing observations, and – after some fearsomely difficult calculations – he predicted where the object should reappear. Sure enough, using Gauss's predictions, astronomers recovered the object later the same year. Named Ceres, it proved to be the largest of a vast swarm of so-called minor planets, or asteroids, orbiting the sun.

While Gauss was hailed for his astonishing achievement, he harboured doubts about the basis of his error formula. Fortunately, it was put on a solid foundation by another discovery, one of far greater significance than the discovery of Ceres.

It was made by another nineteenth-century titan of applied mathematics, Pierre Simon de Laplace (1749–1827). Already celebrated for major discoveries in calculus and celestial mechanics, the brilliant French polymath turned his attention to probability. In 1810, he revealed something about the Bell Curve that even de Moivre and Gauss had missed. In a mind-boggling feat of mathematics, Laplace showed that the roots of the Bell Curve ran far deeper than anyone had suspected, giving the curve huge importance. It was nothing less than a law of nature – one that one could expect to find lurking in a host of phenomena, including some seemingly devoid of all rhyme or reason. Hints of this can be found in the curious ubiquity of bell-shaped curves in the outcome of chance events like coin-tosses. Despite each coin-toss being random and completely independent, when their combined effect is totted up en masse, they somehow conspire to produce the same shape. For example, if 100 people can be persuaded to toss a coin 50 times each and note the total number of heads they see, around a dozen people will get the expected total of 25 heads. All told, around 50 people will get within plus or minus two of this average value. But beyond that, the number of people getting totals farther from the average starts to tail off quite rapidly. Barely a dozen people will get totals more than five away from the average, while just one or two will be unlucky enough to get fewer than 17 or lucky enough to see more than 33. Plotted on a graph, the result will be a bell-shaped curve showing how many people got various total numbers of heads.

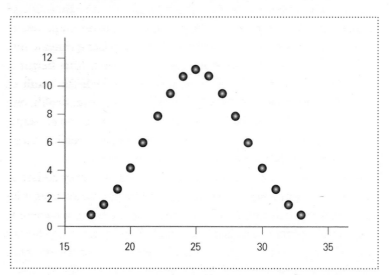

If 100 people toss a coin 50 times, barely a dozen can expect to get exactly 25 heads

Laplace's monumental discovery was that the very same bell curve will describe *any* phenomenon resulting from the combined effect of *any* types of random influence acting independently of each other. Incredibly, we don't even need to know exactly what those influences are or how they behave. Roughly speaking, as long as they're plentiful, of the same type and act independently of each other, their combined effect will produce a bell curve.[4]

If you are struggling to grasp the implications of this, you're in good company: neither Laplace himself nor his contemporaries immediately understood its full significance. It took more than a century for Laplace's discovery to acquire a title reflecting its key role in understanding uncertainty. By rights, it should be called something like the Fundamental Law of Random Influences. In fact, it's known by the stunningly prosaic title of the Central Limit Theorem.

Its applicability is anything but ho-hum, however. Given that many phenomena might reasonably be thought of as the cumulative effect of myriad random influences, one might expect Bell Curves to be ubiquitous. Sure enough, they've been found in everything from the jittery path of gas molecules to the exam grades of students to the heat left over from the Big Bang. The quintessential example is the Bell Curve of human stature. Given that height is the sum total of the length of various bones, each the result of myriad influences from genes and nutrition to general health, one would expect a Bell Curve to appear when the proportion of people with different heights is plotted against the various height ranges. And *voila*! One duly appears.[5]

The Central Limit Theorem does more than supply ballast to lightweight arguments, however. Its stunning generality gives it an almost miraculous ability to cut through complexity. Nowhere is this more apparent than in medical research. To find out whether a new therapy works, clinicians recruit patients and divide them randomly into two groups, one to get the new therapy, the other getting an alternative. This random allocation cuts the risk of either group being abnormal in some way, thus increasing the chances that the results will be representative of the typical future patient. Clearly, it's impossible to account for every quirk of every patient's response, but the Central Limit Theorem makes it unnecessary. As long as these 'unknown unknowns' affect each patient independently, their cumulative effect will be a Bell Curve for each patient group. And if their peaks are sufficiently far apart, it will be hard to dismiss the difference as some lucky fluke.

Clinicians are more aware than most of the presence of the term 'limit' in the theorem's name. This is a reflection of the fact that the theorem strictly holds only in the case of an infinite number of random variables. In reality, it does a pretty good job with relatively small numbers; even so, unless a clinical trial involves enough patients, there's a risk the drug's true impact will be swamped by the 'unknown unknowns'. To deal with this, clinicians flip the theorem around to estimate roughly how many patients they need to include to give a reasonable chance of demonstrating that the therapy actually works – as revealed by two nice, distinct Bell Curves for each group.

The Central Limit Theorem is undoubtedly one of the most powerful tools ever handed to scientists by mathematicians. Its sheer generality is bewitching, and in providing a rigorous underpinning for the Bell Curve, it sparked a revolution in applying mathematics to messy real-world phenomena. But it has also become the poster child for what can go wrong if a mathematical tool is misused, and its 'terms and conditions' ignored. The power of the theorem has led to it becoming embedded in techniques used throughout science, technology, medicine and business. Yet few know of its presence, let alone the risk of breaching the 'Ts & Cs' governing its use. As a result, Laplace's theorem and its progeny have often been pushed too far, resulting in unreliable research findings, nonsensical insights and a central role in the greatest financial crisis of recent times.

The warning signs were apparent even as Laplace struggled to understand the implications of his discovery. In August 1823 a Belgian astronomer named Adolphe Quetelet (1796–1874) walked into the illustrious Paris Observatory on 'one of the most famous short trips in the history of science'.[6] His aim had been to prepare himself for the directorship of a new observatory in Brussels, and in particular to understand how best to extract insight from data. Quetelet met up with many luminaries of science, including

Laplace. But after seeing how the Bell Curve could be used to describe observational errors, Quetelet began pondering more exciting applications. Might the curve be lurking in data on the characteristics of humans?

Quetelet imagined capturing all the essential qualities of humanity via Bell Curves, and of one day unveiling the prototypical human, or, in his phrase, '*l'homme moyen*' – Average Man. And in the years that followed, he began publishing evidence to back this notion. Collecting together data on a host of human traits, Quetelet started to find Bell Curves everywhere, from the chest measurements of soldiers to the propensity to marry or commit crime. Convinced he had found a 'law' of human nature, he began using it to extract insight from data sets. Using the Bell Curve that captured the heights of humans, Quetelet compared the curve for the general male population of France with that of men drafted into the army in 1817. All else being equal, the curves should have been the same, but they weren't: there was a curious 'kink' in the distribution near the height limit for conscription. Quetelet believed his law had revealed that around 2 per cent of men called up for military service had dodged it by lying about their height.

Before long, Quetelet's work on the Bell Curve began to be seen as support for the emerging concept of 'social science', with all kinds of human traits being seen as the sum total of unseen random influences. Quetelet himself believed the ubiquity of the curve was a manifestation of the law of errors as investigated by Gauss and Laplace. To him, the 'Average Man' represented perfection, with deviations away from perfection being the result of 'errors'. But some sought a less metaphysical explanation, and they believed they had found it in Laplace's theorem. For them, the sheer ubiquity of the Bell Curve simply reflected the sheer ubiquity of phenomena resulting from random influences added together. Laplace, it seemed, had built a bridge between the Platonic world of mathematics and the messy world of real life. Who could resist striding across it? Certainly not the Victorian polymath Francis Galton, who more than anyone else became convinced of the universality of the Bell Curve.[7] By 1877 he had begun referring to what mathematicians had variously called the Law of Errors or the Gauss–Laplace Law by an altogether more potent name: the Normal Law. The implication was clear: the Bell Curve reflected the typical behaviour of natural phenomena, the usual state of affairs, the standard way of things. Other influential researchers began to do the same, among them Karl Pearson, one of the founders of modern statistics.

But others were concerned about a dangerous circularity underpinning the belief that the Bell Curve was 'normal'. Among them were the distinguished French mathematician Henri Poincaré and the Nobel Prize-winning physicist Gabriel Lippmann, who noted darkly: 'Everyone believes in it – experimentalists believe it's a mathematical theorem, mathematicians believe it's an empirical fact.'[8]

As we shall see, their concern about such mutually destructive assurance has proved all too prescient.

↑UPSHOT

Of all the laws underlying the behaviour of chance effects, none is more beguiling than Laplace's Central Limit Theorem, and its explanation of the seemingly ubiquitous Bell Curve. But don't be lulled into complacency by textbook talk of the 'Normal Distribution' – because normal it ain't.

The dangers of thinking
everything's Normal

Over its 150-year-plus history, the American investment bank Goldman Sachs has seen it all. Economic booms, financial crashes, stock bubbles, global recessions – whatever the crisis, it has sailed a course through them all. But in August 2007, it crashed into the financial equivalent of a flotilla of icebergs, and had to stuff more than $2 billion into two funds to stop them going underwater. As the bank's Chief Financial Officer, David Viniar was supposed to be the sharp-eyed lookout on the bridge. So how had he missed these leviathans? The account he gave to a reporter that day has become the stuff of legend among the financial cognoscenti: 'We were seeing things that were 25 standard deviation moves, several days in a row.' Which, translated into English means: 'We had some very bad luck.'

Or at least, that's what those fluent in 'quant-speak' took it to mean. This is the language of quantitative analysts, people who, like Viniar, use mathematical models to understand risk and uncertainty in the financial world. These quants carry a lot of baffling stuff around in their heads – including certain key figures that allow them to make sense of fresh financial data in a flash. They all know, for example, that a market move of '1-sigma' has a 68 per cent probability of occurring by chance, and is so common no one loses sleep over it. But a '2-sigma' event has a probability of just 5 per cent, which makes it officially a 'statistically significant' deviation from the usual run of things. Still, stuff happens. It's much harder to be so sanguine about a 4-sigma event; now we're talking odds of around 1 in 16,000 of this happening by chance. You could expect to go your entire career and not experience a day like that. But even the most experienced quants wouldn't have been able to get their heads around Viniar's 25-sigma events. They're so bizarre that even standard formulas for them break down and special measures are needed to get spreadsheets even to display the odds of such events, so low are they.[1] But when they are finally coaxed to give an answer, it's truly astonishing. Viniar and his colleagues were claiming to have been caught out by an event that should occur on average just once every 10^{135} years. That is a figure beyond even astronomical; it a timescale inconceivably longer even than the age of the universe. And according to Viniar, his company had been on the receiving end of *several* such events.

While there's no reason to doubt Viniar's 25-sigma figure, the staggering low probability it implies is troubling. Sure, incredibly rare events can and do happen all the time. But when several all happen together, one has to wonder: is there something wrong with how the chances were worked out? Doing the calculation demands a so-called probability distribution for the event. These come in a host of shapes and sizes, but in finance there's one that everyone turns to, almost without thinking: the Bell Curve. And why not – after all, is it not literally the Normal Distribution?

Concerns about the routine assumption of 'normality' had emerged almost as soon as the concept began to gain currency, over a century before Viniar's statement. In 1901, the pioneering English statistician Karl Pearson had examined some of the claims made for the supposed ubiquity of the Bell Curve, and found the evidence less than compelling. He wrote that 'I can only recognise the occurrence of the normal curve … as a very abnormal phe-nomenon.' By the 1920s, Pearson was lamenting having helped create the illusion that the Bell Curve was 'normal', declaring that the term 'has the disadvantage of leading people to believe that all other distributions are in one sense or another "abnormal"'.[2] He recommended that the assumption of normality be taken only as a first guess in theoretical studies. Yet such qualms were brushed aside as the Bell Curve became not just the first guess, but the only guess. It was just too elegant, the logic for its ubiquity too compelling, the fit to so many data sets too impressive.

But just how well do real-life data fit the Bell Curve? The textbook example is the heights of people, and the recipe simple enough. First take measurements from lots of people, and plot the percentages of them whose heights fall into various bands – say, 5 millimetres apart. The result is a bar chart whose outline forms a pretty smooth Bell Curve looking something like this:[3]

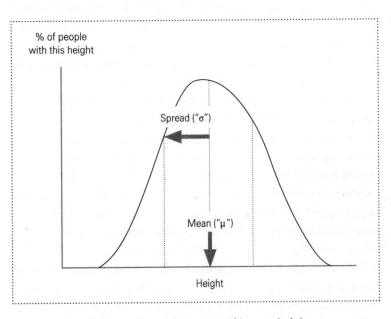

The beautiful, if slightly dented, bell curve of human heights

We also get a real-life demonstration of the power of the normal distribution to summarise a huge mass of data in just two numbers. The first is the mean (that is, average) height, denoted by the Greek letter μ ('mu'). This locates the central peak of the Bell Curve along the horizontal axis. Then there's the standard deviation, denoted by σ ('sigma'), which describes the spread of the Bell Curve. Once we've fitted a Bell Curve to data, knowing just these two numbers is enough to give us a host of information. For example, it's always the case that 95 per cent of the total percentage of the curve lies within almost exactly plus or minus 2σ of the mean. So, for example, if we know that the mean height is 175 centimetres, and the standard deviation is 7.5 centimetres, we know that 95 per cent of people will have heights between around 160 and 190 centimetres. That in turn means that 5 per cent of people lie outside these limits, and as the curve is perfectly symmetrical, we can split them exactly into 2.5 per cent who are shorter than 160 centimetres and 2.5 per cent who are taller than 190 centimetres. We can also flip these calculations around, to ask what percentage of people have heights more than, say, 4 standard deviations ('4 sigma') above the mean. The formula for the Bell Curve shows that around 1 in 16,000 lie beyond 4 sigma of the mean, so by the symmetry of the curve, exactly half this proportion will lie above it. In a country of, say, 100 million people with this distribution of heights, we can expect to find around 3,000 with heights exceeding 205 centimetres.

This is all wonderful stuff, and it's hard not to feel flushed with power. There's just one problem, though – which is apparent if we look more carefully at that real-life Bell Curve. While it's a bell-*shaped* curve, it's not *the* Bell Curve. Laplace's theorem is pretty tough on this point. For any phenomenon presumed to follow its dictates, the Central Limit Theorem tells us that we will get a single, beautiful symmetric curve with graceful tails sloping off to each side. Yet what we've actually got is a squat, stubby curve with a small but distinct dent near its peak. So what's gone wrong? One reason could be that we've not collected enough data to even out all the bumps. That's possible, but not very helpful as we're never going to be *absolutely* certain we've got a perfect Bell Curve, because the theory demands an infinite number of data-points for that. So what else might be causing problems? Dented peaks can be a sign that we've inadvertently mixed up two different popu-lations, with different random influences at work. In the case of human heights, we can at least guess what two of these 'different populations' are: males and females. Sure enough, if we separate the two genders out, we do get better-looking Bell Curves, but they're still less than perfect. OK, so maybe that's because splitting them into just two populations isn't enough – perhaps there are subgroups within subgroups. That makes sense: there will be ethnic background, nutritional status and who knows what else. Now we run into another problem: Laplace's theorem demands that all these different random effects must act independently to give us a real Bell Curve. But is that really plausible? Genes certainly don't act independently of each other, and neither do nutritional influences – and arguing they all exert only additive effects as Laplace's theorem demands is a triumph of hope over experience. In short, the wonder is that the top of the curve looks remotely round and smooth, or its slopes symmetrical.[4]

Some of the early advocates of the 'ubiquitous Bell Curve' theory recognised these challenges. Quetelet tried separating out gender-specific data, and the results were good enough for his studies of '*L'homme moyen*', who by definition lies at the peak of the curve. But some of his contemporaries sought to push on with the whole Bell Curve agenda, to find out what it said about *extremes*. This led them away from the relative safety of the Bell Curve's peak and into its tails. In doing so, they failed to notice – or chose to ignore – that they were at increasing danger of losing touch with reality. Look at any collection of real data on anything, and no matter how much you've got, you'll always be able to find two classes: the biggest ones, and the smallest. Of course, there may well be bigger or smaller ones out there somewhere, perhaps a huge number of them. The trouble is, you don't know; all you can be sure of is that, when you collect your data, you'll always end up with two extremes, with nothing beyond them. But Laplace's beautiful theoretical curve *never* stops. Its tails just continue sloping away for ever, only finally kissing the horizontal axis at infinity. And that has a startling implication for anyone trying to use the Bell Curve to mimic reality. In the case of human height, for example, it means there's a chance – albeit tiny – of finding humans taller than Mount Everest, and some with no height at all, or even *negative* heights. As the probabilities of any of these absurdities coming to pass are so small, it's tempting to treat it as just another quirk, like the dented peak. Yet even as Quetelet and his contemporaries were seeing Bell Curves everywhere, there was actual living proof that the Bell Curve could not be trusted with extremes. It took the startling form of Bud Rogan, the Impossible Man.

Born in Tennessee in the 1860s, by the time of his death in 1905 John William 'Bud' Rogan was 2 metres 67 centimetres tall. His extraordinary height put him well into the right-hand tail of the distribution, making his existence highly improbable. Just how improbable can be estimated using the formula for the Normal Distribution. With typical elegance, it needs just one number to give us the answer: the number of sigma values between Rogan's height and the mean height for his population. Historical records[5] show that for men of his time and background, the mean height was 1 metre 70 centimetres, with a standard deviation of around 7 centimetres. So he towered 97 centimetres above the average man of his time, which is over thirteen standard deviations – or 'sigmas'. Plugging that figure into the relevant formula, it turns out that he was not just a man in a million, or even a billion. He was a man in 10^{44}, or 100 million trillion trillion trillion, which exceeds by a huge factor the roughly 100 billion people who have ever lived. Again, one must never forget that the extremely unusual can and does happen. But, as with Viniar's financial icebergs, we should not expect to see such cases repeatedly. In fact, there are at least seventeen known cases of people with heights similar to that of Rogan, among them Robert Wadlow (1918–1940), who was five centimetres taller and remains the tallest person in recorded history. The lesson is clear: to believe that Everything is Normal is to make assumptions that may not hold good – with consequences that can leave us dumbfounded when dealing with extremes. We must never lose sight of the fact that Laplace's Central Limit Theorem comes with a raft of terms and conditions, and while startlingly relaxed, they can't simply be ignored. Before turning to the Bell Curve for insights, we should always pause to ask whether the data are plausibly the result of the cumulative effect of many variables working more or less independently. The reliability of Laplace's theorem can be undermined by a lack of data – and there's no easy way of telling whether we do have enough. Forced to work in the confines of the messy real world, the theorem rebels by warning of ludicrous possibilities we'll never see – while failing to warn us of extremes we may bump into tomorrow.

Viniar's statement heralded the start of a global financial crisis and recession whose impact will be felt for many years yet. It also prompted huge debate about the assumption of Normality. Not before time: evidence that financial markets do not follow the strictures of the Normal curve has been clear for those with eyes to see for decades.[6] In 2000, the highly regarded British financial mathematician Paul Wilmott tried to warn his fellow quants of the dangers of what they were doing: 'It is clear that a major rethink is desperately required if the world is to avoid a mathematician-led market meltdown … The underlying assumptions in the models, such as the importance of the normal distribution, the elimination of risk, measurable correlations, etc., are incorrect.'[7]

It made not a shred of difference; financial institutions took ever bigger gambles, while holding up the mathematical models as fig leaves to cover their exposure. 'As long as the music is playing, you've got to get up and dance. We're still dancing,' declared Chuck Prince, CEO of Citigroup in early 2007. He wouldn't be dancing much longer. At the time, his bank was the world's largest, with over $40 billion of exposure in the form of Collateralised Debt Obligations (CDOs) – a type of IOU backed ('collateralised') by assets like house mortgages. The attraction of so-called securities like CDOs is the interest rate they pay – typically far higher than a boring government bond. Inevitably, the best interest rates are tied to those IOUs with the highest risk of never being honoured, being collateralised by the dodgiest assets. The challenge lay in deciding whether the interest was worth the risk. Fortunately, credit rating agencies were willing (for a fee) to use sophisticated mathematical models to quantify the dodginess – or 'default risk', in the jargon. But the models weren't very sophisticated at all. They all had Bell Curves built into them – and, worse, they were being used to estimate the risk of extreme events. Ironically for businesses notorious for enforcing 'terms and conditions' with clients, the financial institutions seemed neither to know nor care about those of the Bell Curve. Yet it hardly requires a PhD to suspect that they were likely to be severely breached in the risk models of CDOs. Within months of Prince's cheery statement, the risk models had revealed their inadequacies, and CDOs began to default at catastrophic rates. Citigroup faced bankruptcy, and had to be rescued by a $45 billion bailout from the US government. It was not alone; by early 2008 the global financial crisis had shown that Wilmott's warnings had, if anything, not been dire enough. 'I've been making a big mistake,' he wrote at the time; 'I've been too subtle … Screaming and shouting is needed.' He spelt out his warnings more bluntly, stating that the lack of independence lurking in the models could lead them to 'blow up dramatically'. He advised an immediate cessation in their use. That didn't happen. Securities like CDOs and derivatives like credit default swaps are just too useful – and profitable – to be ignored by financiers. But the quid pro quo has to be that if they are used, they're combined with something more sophisticated than Bell Curves and wishful thinking. Regulators are calling for better models, but to date they don't seem like a vast improvement.[8]

But the real drivers of change are those in the boardrooms of financial behemoths. If we are to avoid a repeat of the recent global meltdown, they must be more knowledgeable about what their quants cook up – and if it goes wrong be compelled to face the music, rather than waltz off to it. There are signs that the message may have begun to get through. When US Treasury markets underwent a 7.5-sigma move in a single day in October 2014, JPMorgan CEO Jamie Dimon told shareholders that such events should happen only once every few billion years. But then he added a telling rider: as the Treasury market has been around for only 200 years or so, 'This would should make you question statistics to begin with.'[9]

It may have taken a calamity to bring it about, but it seems we may finally be moving beyond the Bell Curve.

↑UPSHOT

The Bell Curve comes with 'terms and conditions' that often can't be met in the real world. Sometimes it doesn't matter too much. But if you're using a Bell Curve to predict extremes, watch out: abuse the 'terms and conditions' and you might trigger a catastrophe.

Ugly sisters and evil twins

ake Wobegon, Minnesota, is a very special place. Its most famous son is the best-selling storyteller Garrison Keillor, who has been captivating audiences with monologues about his home town since the 1970s. And, as he delights in explaining at the close of each of his accounts, the town is a place where 'all the women are strong, all the men are good looking, and all the children are above average'. Many may see this as a typically whimsical expression of his pride in the place. Others will see it as a big clue to the truth about Lake Wobegon: it doesn't exist – because such children are impossibilities.

Well, not quite: it's perfectly possible for all Lake Wobegon's children to be, say, above average in some universally defined trait like, say, height or IQ; every sprinter in the 100-metre Olympic final is certainly above average at running. But Keillor is suggesting that all the children are above average *in every way*, and that's a stretch. In fact, if a particular characteristic follows the Bell Curve, then there's only a 50 per cent chance of someone picked at random being above average. The flip-side is also true, and has a scary implication about IQ (which, as it happens, does follow a reasonable Bell Curve): half the people in the country have sub-average intelligence. All this reflects a peculiarity of the Bell Curve: its peak shows not only where the average ('mean') value lies, but also the value of the *median*.

Like the mean, the median is a so-called summary statistic

– a single number which sums up a collection of data. It's often regarded as just some fancy term for the same thing as the average, but it's quite distinct, and often much more informative. For all its familiarity, the average represents something pretty esoteric: it's the best estimate of what you'll get if you pluck one value out of the data at random. That's handy for characteristics like height or IQ, which follow some nice, fairly *symmetric* distribution like the Bell Curve, with as many above the most common value as below. But it can also prove wildly misleading when used with real-life phenomena that don't follow such nice distributions.

In contrast, the median is a pretty robust metric. It's defined as the value which splits the data in half, so that 50 per cent of all the measurements lie below it, and 50 per cent lie above. For data that follow a Bell Curve, the median turns out to be the same as the average value,[1] but the great thing about the median is that it performs its role even if the data *don't* follow the Bell Curve. Indeed, they can follow all kinds of less lovely distributions and still give a well-defined median that divides data evenly into 'high' and 'low'.

That makes the median especially useful if you're suspicious that some characteristic doesn't really follow a Bell Curve. For example, imagine you're applying for a job at a small company of around a dozen people which claims its salaries average around £40,000. That sounds impressive … until you realise that salaries aren't typically distributed according to a Bell Curve, or indeed anything symmetric. Nor is it likely you'll get a salary plucked at random from this range, which is what the average gives. If it's like most companies, the salaries will be highly skewed, with the bulk earning modest amounts, while a handful of fat cats take home a fortune. So unless you're applying to be a fat cat, you should ask to know the *median* salary. The difference can be astounding: in the case of Less4U Ltd (see below) it will reveal that in fact *no one* gets the average figure, because it's rendered meaningless by the huge disparity of the fat cat's salary, and the median is a disappointing £25,000. In general, whenever the median comes out radically *below* the average figure like this, it's a sign the distribution is heavily skewed towards *lower* values – the average being misleadingly inflated by outliers, in this case the salary of one fat cat.

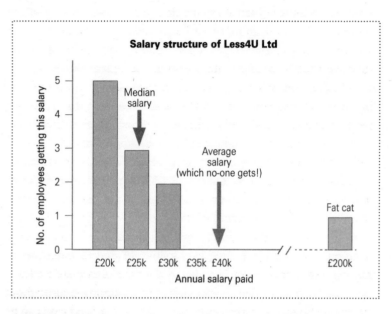

Asking for medians rather than averages can stop fat cats fooling you

Skewed distributions may not look as pretty as Bell Curves, but they're far from rare. Indeed, the world's men provide an excellent example – in the shape of their penises. Or, to be more precise, size: according to research,[2] the average penis length is 13.24 centimetres, but the median value is 13.00 centimetres. This reveals two intriguing facts. First, it shows that the global distribution of penis sizes is skewed towards smaller values, and secondly that most men really do have *below*-average-sized penises. Another example of a skewed distribution concerns driving ability. Many of us claim we're above-average drivers,[3] a belief often dismissed as patently ludicrous; indeed, it has been ascribed to a psychological effect known as *illusory superiority*. But again here we must be wary of falling into the trap of assuming a Normal Distribution. In the UK at least, young drivers are far more likely to be involved in serious accidents, despite making up only a tiny proportion of the total number of drivers.[4] This means the distribution of driving ability is skewed in a direction implying most drivers *are* better than average – though whether it's quite as high a proportion as we drivers believe remains unclear. In general, though, we need to be careful of dismissing seemingly 'stupid' statements of the form 'most X are better/worse than average'. Skewed distributions can pop up anywhere.

They're certainly far from rare in the natural world, appearing in everything from meteorology and ecology to geology. That's partly because real-life phenomena are compelled to lie within decidedly finite ranges. Take heights: according to the Bell Curve, it's possible to have people with zero or even negative heights, but common sense suggests otherwise. Researchers are thus often compelled to tweak their raw data ('log-transform it' is the polite phrase) to reduce the bunching at one extreme and hammer it into something more bell-shaped. This isn't quite the cheat it seems: it amounts to claiming the phenomena are due to independent random influences *multiplying* together, rather than merely adding up.[5] And such 'multiplicative' phenomena are common throughout the life sciences, chemistry and physics. Indeed, a good case can be made for stripping the Bell Curve of its misleading moniker of the 'Normal Distribution', and conferring the title instead on its less famous logarithmic relation.[6] With its lack of symmetry, this 'ugly sister' of the beautiful Bell Curve might lack aesthetic appeal, but it may better reflect the ugly world we live in (see overleaf).

Yet those who mindlessly use the Bell Curve run the risk of witnessing something far nastier than mere loss of symmetry. They can find themselves confronted by the truly monstrous consequences of the failure of Laplace's Central Limit Theorem. Fittingly, the first glimpse of them came via a curve first studied by eighteenth-century mathematicians known as the Witch of Agnesi. Quite how it came by that name isn't clear, but it seems appropriate given the demonic effects predicted if it is ever found lurking in data. At first sight, it appears to be just like the Bell Curve: a central peak with graceful slopes descending symmetrically to either side. But there's something not quite the same – as becomes clear when a Bell Curve is projected onto it (see overleaf).

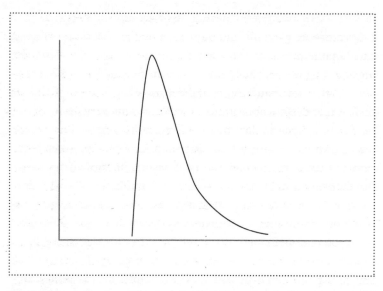

The log-Normal curve: uglier than the Bell Curve, but perhaps more useful

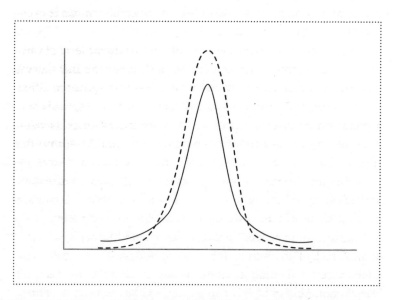

The Witch of Agnesi (solid line) is trying to fool you it's Normal (dotted)

The Witch of Agnesi's peak is sharper, more pointed, but its slopes are more graceful, and more reluctant to fade away to either side.[7] Mathematicians call such curves 'leptokurtotic', from the Greek for 'lightly arching', but those who've run into it in real life have a rather less flattering name: 'fat tailed'. It's symptomatic of the fact that despite appearances there's nothing very pretty about the Witch of Agnesi. Data that conform to its shape follow what's now known as a Cauchy Distribution, after a prolific nineteenth-century French mathematician. And despite its similarities to the Bell Curve and following a far simpler formula, the Cauchy Distribution is a nest of mathematical vipers. First, data that conform to it refuse to possess an average value. Sure, it's possible to take, say, 1,000 data-points and work out their average by adding them up and dividing by 1,000, but the result will be *meaningless*. The next data-point could be so wildly different from everything else that it utterly changes the average. You might be chugging along with data in the 10s and 100s when suddenly – kerpow! – a value of 51,319 suddenly appears. Unlike data following a Bell Curve, where adding more data gives a better estimate of the average value, adding more Cauchy data makes no difference: all you get is ever-changing values.

It's the same story with any attempt to estimate the level of variability, as captured by the standard deviation. For the Bell Curve, the standard deviation is reflected by the degree of spread to either side of the central peak. The Cauchy curve clearly also spreads out, so the standard deviation isn't zero. But try and estimate its value using 100, 1,000 or a trillion data-points, and you'll run into the same problem as with the average: the results just zoom all over the place. In other words, the average ('mean') and standard deviation of the Cauchy aren't big, small or somewhere in between. Despite what the shape of the curve suggests, they simply *don't exist*.

Statistics and probability textbooks typically devote little space to the Cauchy Distribution. If it's mentioned at all, it's usually portrayed as just some mathematical freak that looks like the Normal Distribution but isn't.[8] But that's precisely the reason the Cauchy Distribution should be far better known: it's the poster child for the dangers of assuming Everything's Normal. And nowhere is that more apparent than when trying to estimate the chances of freak-ish outcomes. These are, by definition, unusual and thus lie in the very-low-probability tails far from the central peaks of the Bell Curve or Cauchy Distribution. But one glance at the two curves superimposed on each other above shows that they're not going to give the same answers. The thicker 'tails' of the Cauchy Dis-tribution suggest it's going to ascribe a higher chance of witness-ing freakish outcomes than the Bell Curve. But just how much higher requires calculation, the results of which are given in the table below:

Bell Curve probability: 1 chance in ...	Equivalent Cauchy probability: 1 chance in ...	Bell Curve underestimates chances by factor of ...
20	7	3
100	9	11
1,000	11	91
1 million	16	62,500
1 billion	19	53 million
1 trillion	23	43 billion

If you trust the Bell Curve to assess rare events, prepare for shocks

The differences between the predictions are truly shocking, especially given the apparent similarities of the two curves. And that shows the dangers of blithely assuming that data that fit some-thing *like* a Bell Curve really is normal. This is especially true for rare events. For example, an event expected on average once in every billion years assuming it's 'normal' could pop up once in nineteen years if it actually follows the Cauchy Distribution. Sud-denly it's no surprise that the likes of JPMorgan's Jamie Dimon has lived to see a one-in-a-billion market move. Even David Viniar's

2007 report of having experienced in a few days events that should never have happened in the history of the universe no longer seems so outlandish.[9] Or at least, it doesn't if one believes that the Cauchy Distribution really could apply to such events. But is that really plausible? Can events in real life really follow something as whacky as the Cauchy Distribution, with its bizarre reluctance to give even average values?

Given the fortunes riding on it, it's no surprise that researchers have been trying for decades to fit distributions to financial data. And given the penchant for seeing the Bell Curve everywhere, early studies claimed stock price movements did indeed follow its dictates. Yet as long ago as the mid-1960s it was clear this was wishful thinking. In a celebrated PhD thesis published while he was still in his twenties, the American economist and future Nobellist Eugene Fama showed that there are just too many extreme stock price swings. That gave the distribution a central peak that is spikier and its tails thicker than expected from a Bell Curve[10] – in other words, rather like the Cauchy Distribution. But Fama found it was more interesting than that. The best fit came from using curves belonging to a whole family of distributions of which the Cauchy and the Bell Curve are just special cases. Known enigmatically as the Lévy-stable distributions,[11] they can be as benign as the Bell Curve or as crazy as the Cauchy.[12] Fama found that stock price moves have a distribution somewhere in between.

What wasn't clear was why. Clearly, their behaviour must violate at least one of the 'terms and conditions' of the Central Limit Theorem, which underpins the Bell Curve – and the obvious candidate was independence. After all, everyone knows that investors are like sheep, all buying 'hot tips' or selling 'dogs'. Yet Fama found that the stock price on any given day was more or less independent of its value up to sixteen days previously. So if the assumption of independence was OK, what else could have gone wrong? Fama found the clue in the sheer violence of stock market moves. Like the Cauchy Distribution, they have a pathologically large standard deviation, and can be both huge and sudden. Such behaviour cannot be countenanced by the Central Limit Theorem, and its Bell Curve gets distorted into something pointier, thicker tailed and altogether more dangerous.

This should come as no surprise to anyone who has experienced the financial roller-coaster of the last few decades. What should scandalise us all is that it was all known about over half a century ago. Academics like Fama had shown that while prices on any particular day might follow the dictates of the Bell Curve, they're still capable of moving with terrifying abruptness. As such, trusting the Bell Curve to estimate the risk of a given loss is itself extraordinarily risky, almost criminally irresponsible. Yet despite all this, Bell Curves remained embedded in the business of estimating risk, even in the financial sector.

The Cauchy Distribution is the Bell Curve's evil twin, able to pass itself off as its far more benign sibling with its neat peak and graceful tails, but capable of behaving very badly. But it's not alone. As well as its close relatives in the Lévy stable family, the Cauchy curve's nastiest traits are shared by so-called power law distributions – which have been found lurking in a host of real-life phenomena, from earthquakes to forest fires to personal wealth. Mathematically, power laws are far simpler than the Bell Curve, but do the same job of linking the size of a phenomenon to its prevalence (see graph opposite). Their origins seem to be as varied as the phenomena they describe,[13] but they all share the same basic appearance: no central peak, but more of a cliff-edge which plunges down and then stretches out in a long tail reflecting the key feature of the phenomena they describe: bigger means rarer. Take earthquakes: while most are too feeble even to notice, some are unnerving – and a few are devastating. Historical records have allowed seismologists to pin this down more precisely, leading them to the so-called Gutenberg–Richter relation. This shows there are ten times fewer quakes with Richter magnitudes between 6 and 7 as between 5 and 6, and ten times fewer still between 7 and 8. So dramatic a decline is typical of a power law, and in this case it's a pretty simple and well-behaved one (see graph overleaf).

It does at least allow stable values for the average sizes of earthquakes to be extracted from data – which is more than you can

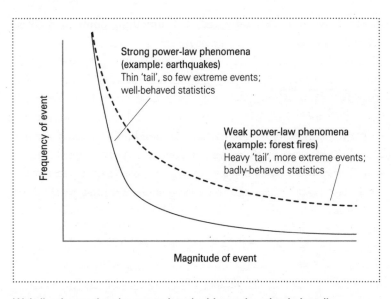

Weirdly, the weaker the power law the bigger the sting in its tail

expect from a Cauchy curve. But not all power law phenomena are so benign: solar flares, forest fires and human conflict have all been shown to follow power law distributions for which neither average sizes nor even plausible ranges can be reliably estimated. These power laws are just as reluctant as the Cauchy curve to play statistical ball with us. And that has serious practical consequences. For instance, how can one be sure of tackling the risk of, say, major forest fires when even their average size is so hard to estimate? The existence of power curves also threatens the reliability of insights into phenomena mistakenly believed to follow Bell Curves.[14] Researchers run the risk of blithely calculating such basic statistics as averages, unaware that power laws driving the data may make such statistics meaningless. As we'll see, they can also trash standard methods of analysing data and finding patterns, and undermine attempts to replicate claims based on them.

In short, these pathological distributions have the power to undermine the very methods of science itself. As such, they present us with a fundamental challenge: do we accept their existence, and learn to live and work with them – or should we continue to trust models of reality that are simple, elegant and wrong?

⬆UPSHOT

The real world harbours a host of phenomena that look entirely normal, but are anything but. Worse still, data on these mathematical monsters can make them look completely benign. Yet unless they're spotted and dealt with carefully, they can make a mockery of our attempts to understand them.

Going to extremes

The idea of pinning the blame for a global crisis on one person rarely makes sense. But in the case of the financial meltdown of 2007/08, one name has come up more than any other: Alan Greenspan. From 1987 until just a few months before the crisis erupted, Greenspan was chairman of the US Federal Reserve – in other words, head of the central banking system of the world's biggest economy. During that time, say his critics, he instituted a regime of ever-looser financial regulation, driven by an almost religious belief in the benefits of free markets. The result was unfettered greed, insane levels of leverage and risk-taking, and a multi-trillion-dollar disaster.

There's no shortage of evidence against Greenspan – not least his own *mea culpa* before a Congressional committee in 2008, in which he admitted to being in 'a state of shocked disbelief' about what had happened on his watch. Yet he deserves credit for being among the first to express concern about the dangerous assumptions lurking in the risk models used in finance. Speaking at a conference of central bankers in 1995, Greenspan warned of the 'inappropriate use' of the Bell Curve, with its tendency to underestimate the chances of extraordinary events. His audience was starting to get uncomfortably familiar with those. In February of that year, the world's most famous merchant bank, Barings, had collapsed after losing over £800 million (equivalent to around £1.5 billion today) through the dealing of a single trader named Nick

Leeson. Then the Daiwa Bank of Japan discovered a similarly vast hole in its accounts through the activities of another rogue trader. For Greenspan, the lesson was clear: central banks had to see themselves as insurance companies, able to provide coverage even in the event of catastrophes. And that, Greenspan implied, meant making more use of a mathematical toolkit increasingly relied on by the insurance industry to impressive effect: Extreme Value Theory (EVT).

The idea that extremes were more than just outliers of familiar distributions had been recognised in the early eighteenth century by one of the pioneers of probability theory, Nicolaus Bernoulli. Yet despite their obvious importance, it took another 200 years for the theory behind them to emerge. In the 1920s, the ever-brilliant Ronald Fisher, together with his former student Leonard Tippett, proved that extreme events follow their own special distributions.[1] These were later combined into a single formula, known as the Generalised Extreme Value (GEV) distribution, whose shape can be tuned using data about the extreme events. The resulting curves are mathematically somewhat weird, but still reflect the common-sense idea that the more extreme the event, the less likely it is. Crucially, however, the detailed predictions could be radically different from those emerging from the Bell Curve.

With business models constantly under threat from extreme events, insurance companies have been assiduous students of EVT. For years, actuaries gauged the likely risk posed by various forms of disaster using empirical rules of thumb such as the '20–80' rule, which states that 20 per cent of the severe events account for over 80 per cent of the total payout.[2] In the mid-1990s financial mathematician Paul Embrechts and colleagues at the Swiss Federal Institute of Technology (ETH) in Zurich decided to check the validity of such rules with EVT. They found that the '20–80' rule does work well for many insurance sectors – but when it fails, it does so very badly. Using EVT to study past data on claims, the team found that a '0.1–95' rule applies to hurricane damage. In other words, while all hurricanes are a potential challenge, the real threat comes from the one-in-a-thousand storm, which can devour 95 per cent of all the insurance cover in one gulp. Such discoveries allowed insurance companies to optimise their risk coverage, broadening the range of threats they can cover at sensible premiums, benefiting both themselves and their clients.

EVT is now also used to protect those whose very lives are put at risk by such natural calamities. One nation has actually bet its future on the theory's predictions. In February 1953, a huge storm-surge struck the North Sea coast of Europe. The resulting floods killed over 2,500, including 1,800 in the Netherlands, whose centuries-old sea defences were overwhelmed. Determined to prevent a repetition for generations to come, the Dutch government set up a panel of experts to design sea defences that could meet the standard without bankrupting the country. The panel estimated that coastal defences around five metres above sea level would suffice. But could the figure be trusted? Records showed that the 1953 event was not exceptional: severe floods have struck the Netherlands dozens of times over the previous millennium. On All Saints' Day, 1 November 1570, the country was devastated by a storm surge of over four metres – more than fifteen centimetres above even the 1953 event – resulting in tens of thousands of deaths. The concerns led the Dutch government to commission a team led by EVT expert Laurens de Haan at Erasmus University, Rotterdam, to assess the 5-metre standard. Using historical flood data, the team established the EVT curve that accounted for past extreme floods – and then extrapolated it into the future. They found that the original recommendations should hold good for centuries.

Whether they will or not remains to be seen; as we've discovered, it's never wise to place blind faith in mathematical models, no matter how sophisticated they seem. And there are certainly grounds for concern about the reliability of EVT, because – just as with the Bell Curve – its impressive powers come with a long list of 'terms and conditions'. A key issue concerns the very data used to establish the best EVT curve for the job. As its name suggests, we need examples of extreme cases – but what counts as 'extreme'? Trawling through historical records, some kind of threshold needs to be set, but where? Setting it too low will let in too many ho-hum cases, making the curve inaccurate; too high a threshold will make the data set so thin that the curve becomes fuzzy and imprecise. Then there's the problem that blights the use of the Bell Curve: what's driving the observed data – and are these influences independent and unchanging? Given the evidence of climate change over the centuries, those would seem to be questionable assumptions to make about hurricanes, floods and storms.

Nor is EVT free from that most bizarre yet pernicious of statistical blights: unstable averages and ranges. As with power laws and Cauchy-like curves, some types of EVT distribution are deeply wayward. Research using real-life data about extreme losses incurred by banks has found that the resulting curves often fail to have well-defined ranges or average values.[3] That makes risk estimates very unstable, the addition of just a few more data-points utterly changing both the risk figure and the size of the likely hit.[4]

Clearly, it's not going to be easy to take up Greenspan's proposal and make use of EVT in financial models. Efforts are being made to solve these problems – spurred on by the fact that while EVT is still a work in progress, it's more likely to err on the side of caution than the Bell Curve. The biggest barrier to its acceptance may be the financial institutions themselves. Following the crash, they are now obliged by regulators to carry reserves able to cope when deals and loans go bad, so they never need bailing out again. Working out the size of these reserves is a tough challenge in risk modelling. But it's clear that if it's done using EVT, the reserves are likely to prove substantially bigger than those demanded using the Bell Curve.[5] The trouble is, banks aren't keen to have huge sums sitting around in vaults for a rainy day – and regulators have allowed them to choose how they do their sums.

Will they decide not to take the risk of being caught out again, and spurn the enticingly slender tails of the Bell Curve for the fat and pricey tails of EVT? Given that we've had at least five major financial crises since Greenspan's 1995 suggestion, one can only say: don't bank on it.

Extreme lives – and losing streaks

Since the 1950s, typical life expectancy has risen from around 45 to over 70 worldwide, and now exceeds 80 in many developed countries. This trend clearly can't continue for ever, but where will it stop? Can what we know about human longevity be used to estimate the maximum possible human lifespan? At Erasmus University, Rotterdam, Laurens de Haan and his colleagues examined the lifespan records for the 'oldest old', and then applied EVT to extrapolate to the ultimate human lifespan. They ended up with a figure of around 124 years.[6] At the time, the oldest person ever recorded was still alive: Jeanne Calment of Arles, France, who recalled meeting Vincent Van Gogh at the age of thirteen. She died in 1997 at the age of 122 – just two years short of the upper bound set using EVT, which currently looks safe for some years to come.

While the theory behind EVT is complex, a simple version of it works for one of the most irksome extremes in life: long losing streaks. The resulting formula[7] has some surprising implications. For example, if we toss a coin 50 times, we should prepare to see streaks of heads (or tails) of around five on the trot, plus or minus around two. That's far longer than most people expect – and helps put losing streaks into perspective. It also casts some light on a notorious losing streak encountered by the British horse racing tipster Tom Segal of the *Racing Post*. Using the pen-name Pricewise, Segal has a reputation for recommending unfancied horses at relatively long odds. Such horses are unlikely to win very often, but win big when they do. In 2011, Segal hit a streak of 26 consecutive bad tips – prompting many of his followers to worry he'd lost his touch. Yet EVT shows that for the long-odds tips Segal gives, a streak of 32 losing bets on the bounce would be entirely normal over the course of a year. Sure enough, the bad run ended a few weeks later, and Segal went on to produce an impressive 20 per cent return on investment for those who kept the faith.

↑UPSHOT

In a world assailed by extremes from freakish weather to financial upheaval, Extreme Value Theory can turn historical records into insights about how bad things might get. Assuming the future will be like the past is risky – but if you think that's dangerous, try guesswork.

See a Nicolas Cage movie and die

All scientists want to make discoveries, ones that change our view of life, the universe and even the nature of reality. Most have to settle for some insight that just makes people sit up and take notice. By that standard, Tyler Vigen has been brilliantly successful. His discoveries have been reported worldwide, and are breathtaking in their variety, unexpectedness and sheer number. To date, he's uncovered tens of thousands of mind-boggling insights, and he hasn't stopped yet. Or, to be more accurate, his computer hasn't.

For Vigen isn't doing the discovering; he leaves that to his computer, which he's programmed to do exactly what scientists have been doing for decades: scouring data to find out how one variable changes with the other. It's a technique that has led scientists to a host of discoveries, from links between radiation exposure and cancer risk to the connection between the properties of stars and the expansion of the cosmos. Vigen's computer applies the same methods, analysing the data in search of 'highly correlated' variables. That is, it seeks out random data sets and applies a formula that spits out so-called 'correlation coefficients'. These can range from +1 – where high values of one variable correspond to high values of the other – through zero, where there's no pattern, to −1, where high values of one correspond to low values of the other and vice versa (see below).[1]

Vigen's computer seeks out data sets which produce correlation coefficients close to these extremes when paired against each other.

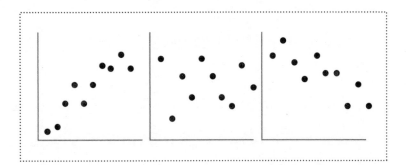

Three degrees of correlation: +0.85, 0.0, –0.85. All can be important –
or twaddle

That's because they're what you expect to find if there really is a
strong connection between two variables. In contrast, correlation
coefficients near zero are symptomatic of a lack of any relationship, and thus nothing exciting going on. By automating the whole
process, Vigen has thus created a Discovery Engine.

What it's discovering should certainly change our view – not
of reality, but of the reliability of many headline-grabbing claims
based on this same technique. Vigen isn't a scientist; at the time
of writing, he's a grad student in law at Harvard. But by letting
his correlation-hunting computer loose on the data-rich pastures
of the web and posting the results on his website, he's providing
a constant reminder of the dangers of mindlessly applying one of
the most popular but misused concepts in science. Since he set it
running, Vigen's computer has found a host of utterly crazy correlations. Taken at face value, they suggest that Nicolas Cage should
be stopped from releasing movies, as they're linked with deaths in
swimming pools (correlation coefficient +0.67), and that America
should ban the import of Japanese cars, as they're associated with
suicide by car crash (correlation coefficient +0.94).

Among the top tips revealed by Vigen's computer is that it's
not a good idea to eat cheese last thing at night, as per capita consumption of the stuff is strongly correlated with death by becoming tangled in bedsheets (+0.97). If you're having problems in
your relationships, you may also want to consider relocating to

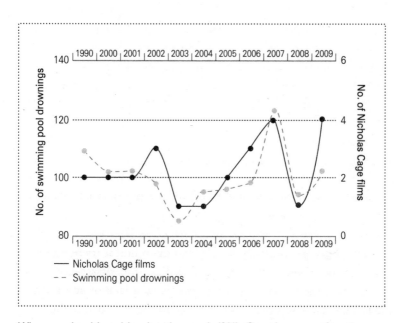

Why you should avoid swimming pools if Nic Cage has a movie out

somewhere that consumes relatively little margarine: Vigen's computer has revealed that per capita consumption of the stuff is very highly correlated to divorce rates – in the US state of Maine, anyhow.

All very amusing, and it's no surprise that Vigen's website reporting these 'discoveries' has had over five million hits. After all, they're eerily reminiscent of the discoveries we encounter so often in the media, accompanied by phrases like '... according to scientists'. It would be nice to think such nonsense couldn't possibly get traction among serious scientists, but Vigen has got an even more important lesson for us on this. Most of his 'discoveries' are statistically significant, the standard litmus test used in research to assess whether a finding is more than just a meaningless fluke.[2] As such, to keep this stuff out of the serious research literature, the techniques relied on by most research workers are too feeble. We have to look beyond them.

Sheer implausibility is the most obvious way. Nothing other

than their implausibility is preventing most of the correlations being taken seriously (e.g. US oil imports from Norway and drivers killed by trains – which has an extremely highly significant correlation coefficient of +0.96). Other correlations crash and burn the moment one looks at the hard figures behind the raw figures. Take the lethality of Nic Cage's movies. He's a very hard-working guy, and has appeared in several films each year for over a decade, but even he struggled to make more than three a year. In other words, his output has been pretty constant. And in that he's a match for the actions of the Grim Reaper around US swimming pools. During the decade of data used by Vigen's computer to find a correlation, there were around a hundred fatalities each year, but never fewer than 85 or more than 123. But by chance, these peaks occurred in the two years where Nic also did his least and most numbers of films. Because the data set is so small, the coincidence of these two extreme sets of points overwhelms the feeble evidence in the other more or less constant values – and we end up thinking Nic and the Grim Reaper are a double-act (with, as it happens, a spookily appropriate correlation coefficient of +0.666). Such 'outliers' are notorious for making and breaking correlations when there's little data to go on. They're often treated as the spawn of 'experimental error' or some other blunder and simply rubbed out in a process euphemistically called 'data cleaning'. In the case of the Cage/ drowning data sets, such cleaning halves the correlation coefficient, which also becomes non-significant. Yet in real scientific research, justifying such elimination is often not so simple. Outliers can be entirely genuine if one is dealing with phenomena with power law behaviour, such as some weather phenomena or economic factors.[3]

Nic is clearly off the hook, but not all of Vigen's 'discoveries' can be laughed off so easily. Can we be sure, for example, that there's really nothing in the correlation between total revenue generated by US golf courses and the amount of money Americans spend on spectator sports (+0.95)? Maybe it's a reflection of the fact that people who've watched golf want to have a go themselves. Or maybe people who play golf are keen on sport generally? The simple strength of the correlation doesn't tell us how or even if the relationship is genuine; as the old saying puts it, correlation isn't causation. Nor does statistical significance say anything about the real 'significance' of a correlation, despite what many researchers seem to think. That, remember, merely measures the chances of getting at least as impressive a correlation *assuming* it's really just a fluke; it says nothing about whether that assumption is actually true. And as we've seen many times, answering that question requires Bayesian methods – and here they bring the extra advantage of allowing us to factor in our prior beliefs about the correlation. In principle, that can help give us a handle on the chances of the correlation being a fluke. Yet even that's tricky, because the correlation might indeed be real, credible, but still a red herring. It might be the product of a hidden 'confounder' – some intermediary that links the two variables that are themselves unconnected. Cases of severe sunburn are doubtless significantly correlated with sales of suntan lotion – and also ice cream and cold drinks. Does that mean they cause sunburn? Of course not. There's a not-so-hidden confounder connecting them all: the sun.

That said, the results of confounding can be amusing. No one knows quite when or how the idea that storks deliver babies got going, but it has acquired legendary status among statisticians, several studies having uncovered a strong and statistically significant correlation between stork populations and births in various countries. One potential explanation is the confounding factor of land area – which is correlated to both stork populations and birth-rates.[4] The effects of confounding aren't always so entertaining, however. Unless identified and corrected for, they can end up driving public policy. Smoking marijuana has been linked to a host of health risks, and even those who've never touched the stuff know it turns you into a halfwit. Confirmation came in 2012, in a study published in a respected journal that found a clear link between cannabis dependency over time and loss of IQ.[5] Aware of the need to avoid being fooled by confounders, the researchers took into account factors like the use of alcohol and hard drugs, but the effect remained: those who had acquired a habit in their teens and became heavy and persistent users had lost eight IQ points by their late thirties. But wait

– don't people get dimmer over time anyway? Possibly, and the researchers had this covered, with a comparison of similarly aged people who'd never used cannabis (oddly their IQ actually went up very slightly). Despite all this, however, the researchers ran into the usual problem in dealing with confounders: the more of them are stripped out, the more data-points end up excluded from the final analysis. Having started with over a thousand people in the original study group, the researchers ended up with just a few dozen free of the confounding influence of alcohol and hard drug abuse. And as the team admitted, these are hardly the only possible confounders. Even so, by confirming 'what everyone knows' about long-term stoners, the study got a lot of media coverage.

Yet within weeks its conclusions were being challenged for having failed to take into account other confounders. One is an intriguing phenomenon involving rising IQ test scores noted in many countries since the 1930s. Known as the Flynn Effect, the reason why people living today are so much 'smarter' than their grandparents (or at least, do better in IQ tests) is still debated, but one possibility – backed by the effect's eponymous discoverer – is that we increasingly live in environments rich in IQ-test-like tasks, and those who are especially good at them find themselves in situations giving them even more challenges, boosting the effect further. Whatever the explanation, the Flynn Effect clearly needs to be taken into account in any study focusing on IQ changes over time – and when applied to the cannabis–IQ study, it easily accounted for the supposed effect of long-term cannabis use.[6] So can stoners just kick back and carry on puffing away? Not quite, because the Flynn Effect is only a potential confounder rather than a proven one. What is beyond doubt, however, is the vulnerability of correlation studies to confounders – and the need to continue looking for them even when we've got the 'right' answer. That's especially important in studies of controversial but common sources of risk such as passive smoking, which are themselves confounders in other studies.[7]

All this might give the impression that correlations are tricky things that lay traps for the unwary. Yet there are some warning signs that should always ring alarm bells. The first is whether the raw data has been bundled up to make it look neater than it really is. One obvious way to do this is to take a whole load of measurements, take their average and correlate those. The averaging reduces all the messy scattered data-points into nice, neat points. The result can be a seemingly much more impressive level of correlation – as many researchers in the 'softer' sciences have noted. In one textbook example,[8] the correlation between the average educational level and income for men aged 25–54 for each of the states of America was found to be +0.64, showing the importance of staying in school. But when the analysis was repeated at the level of individuals using census data, the resulting variability drove the correlation down to around +0.44.

This 'data cleaning' trick is especially deceptive where the amount of scatter violates one of the foundations of simple correlation theory: that the amount of variability stays pretty constant. For example, raw data may have come from different sources of varying quality, or there may just be fewer data-points in some places than others. The result is more uncertainty, and potentially misleading correlations. Research into scary health risks is particularly vulnerable to this. There are often plenty of people getting low exposure, but relatively few getting high levels, increasing the uncertainty and the level of scatter as the level of exposure increases.

Scatter can also emerge from the variables themselves. There may be some unknown factor at work, or maybe one of the variables simply doesn't have a well-defined variance; as we've seen, there are plenty of those in nature. And it's possible that several of these effects might be happening simultaneously. Whatever: the upshot is that simple ways of trying to mask the problem via nice, neat averages might make for more convincing graphs, but the resulting correlations and other inferences can be hopelessly misleading.

Warnings about how simple tweaks to data can undermine the reliability of correlations have been sounded ever since they were first used. Indeed, the very mathematician who developed the basic theory, Karl Pearson, cautioned researchers about correlations based on *rates*, such as X 'per thousand people' or 'per month'.

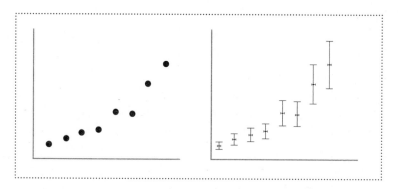

Correlated data: the neat presentation version – and the messy, uncertain raw stuff

These are often used in business as well as academic research with the aim of putting everything 'on the same footing', but both theoretical and empirical research has shown Pearson's fears are well founded[9] – which is pretty worrying, given the plethora of 'relationships' claimed on the basis of rate-based correlations.

Over half a century ago, the celebrated statistician Jerzy Neyman declared that 'Spurious correlations have been ruining empirical statistical research from times immemorial.' Plausibility – or the lack of it – plus the old dictum that 'correlation is not causation' can save us from reading too much into very little. But we shouldn't forget that the flip-side is also true: lack of correlation doesn't necessarily imply lack of a genuine relationship. After all, the 'Ts & Cs' of the theory of simple correlation assume the relationship is linear, and there are plenty that aren't. Take a look at the figure opposite.

At first glance, there seems to be some sort of relationship here – but the simple theory of correlation tells us there isn't: the correlation coefficient is just 0.36, and it has a hopelessly non-significant p-value of 0.25. But those two numbers are only really telling us two things. First, if there is a relationship, it's not a simple straight line … which you might possibly have twigged by just eyeballing the graph. Then the p-value is saying that the chances of getting at least as poor a straight line as we did by fluke alone are quite high – another unhelpful insight. Even so, if – like all too many

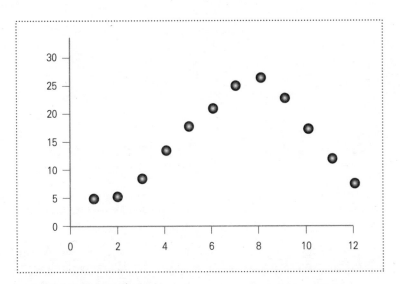

It's clear something's going on here – unless you mindlessly use correlation analysis

users of statistical methods – we ignore the limitations of simple correlation theory (if we ever knew them), and misinterpret the high p-value as being 'the chances the result is just a fluke', then it's case closed: there's nothing going on here at all. All of which defies common sense – and ensures we'll miss a key insight about when to visit Japan, as the points show the link between the month and the typical temperature in Tokyo in Celsius, which is of course both real and significant – in every sense of that much-abused word.[10]

↑UPSHOT

Correlations are like coincidences: we'd take them far less seriously if we were more aware of just how easily it is to find them. Powerful methods for measuring correlation exist, but they can all prove misleading if we insist 'there's got to be something in this pattern'.

We've got to draw the line somewhere

No one has put the laws of physics to more impressive use than the US space agency NASA. In January 2006, it fired an object the size of a grand piano towards a moving target 4.5 billion kilometres away at around 50,000 kilometres an hour. Nine years later, the New Horizons probe zipped past Pluto 72 seconds ahead of schedule in a close encounter equivalent to hitting a hole-in-one from a distance of 30 kilometres. NASA can pull off such feats because its scientists and engineers are very smart and truly have little to worry about: with just the vacuum of space between them and their target, mission planners can get away with using the law of gravity plus a few tweaks to make predictions of stunning reliability. They can declare with near-complete confidence that if they successfully launch on a specific date at this speed on this trajectory, they will end up at that point on that date.

Back on Planet Earth, things aren't so simple, but the same question arises in myriad contexts: if this happens, what will follow? If the prevalence of greenhouse gases continues to rise, what will happen to global temperatures? If we charge more for our product, what will be the impact on sales? If this, then what?

As it happens, by far the most widely used method for finding out was invented for astronomical purposes over two hundred years ago. The German polymath Carl Gauss – he of Bell Curve fame – appears to have used it to help (re)discover the first known asteroid, Ceres, in 1801. It's known as the method of least squares or, barely less opaquely, linear regression. In essence, it just puts a straight line through messy data, but not just any straight line; it finds the best-fitting one. The exact definition of 'best' here is a bit technical,[1] but in essence it means it does a mathematically precise job of what you'd do if asked to put a line as close to as many data-points as possible:

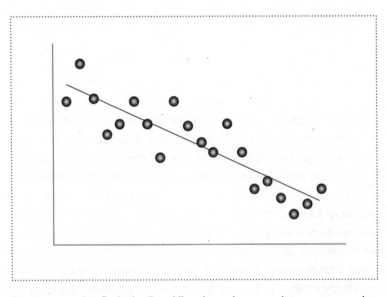

Linear regression finds the 'best' line through messy data – up to a point

Armed with this best-fit 'regression line' extracted from our data (any spreadsheet will do it), all kinds of things become possible. We can use the line to make up for gaps in the data. We can use the slope of the line to gauge the impact of changing one variable on the other. We can see when and where one or other variable becomes zero. Hell, we can even use the regression line to go *beyond* our data. Imagine: we could have financial market data with stock prices at different times, use linear regression to get the best-fit line and then *predict* what the price would be tomorrow, even next week or months ahead. We could be *rich*.

If you've got this far into the book, you'll have twigged there may be something wrong with all this. What you may not have realised is just how many otherwise smart people haven't twigged it. The most basic problem with reaching for linear regression to find relationships in data is the same one we ran into with correlation: the very idea that there's any relationship at all. Imputing cause from correlation is risky even when done thoughtfully. When done mindlessly the results are at best laughable. A few clicks on a spreadsheet allows linear regression to deduce Cage's Law in all its mathematical finery:

No. of drownings = 5.8 × no. of Nic Cage films + 87

We can top it off with an impressively high correlation coefficient (+0.67), adding as a final flourish the fact that the correlation is statistically significant (p = 0.025). Taken seriously, this regression equation is telling us that each new film featuring Mr Cage causes another six deaths by drowning. But surely no one would take it seriously; the law is patent nonsense, because ... well, it just is. And therein lies the problem with regression analysis: it says nothing about whether it even made sense to try it. We are still waiting for spreadsheet software which can spot ill-advised attempts to find relationships in data and flash up the best-fit line along with the message 'You're just messing about, right?'.

Slightly higher up the scale of sophistication is assuming it's OK to fit a simple straight line to data that really reflects something more complex. Again, there's no point looking for advice from our software. Like Dr Frankenstein's faithful assistant Igor, it will mindlessly do whatever we ask, no matter how gruesome the outcome. Even if the data-points followed the outline of a banana in a bun, linear regression will put a best-fit line through them.

It will even indulge us in our desire to act like gods and foretell the future. And why shouldn't it? After all, if we can use linear regression to show how, say, product sales vary with advertising spend, why shouldn't we be able to do the same to predict, say, sales over time? No reason at all – except that time isn't just another variable. It has a nasty habit of linking things together. And that, in turn, raises that old problem: breaching the little-read 'terms and conditions' of the mathematical kit we're using.

Buried in the little-read 'Ts & Cs' of linear regression is a demand that there's no pattern to the errors made by the 'best-fit' line as its passes through the data-points. As ever, that just sounds tiresome and complicated, but as so often it's also crucial, as it's precisely what can appear in data covering a span of time. Everything from business cycles and seasonal effects to simple momentum can create ties between data-points, and the resulting 'autocorrelation' can make a mockery of any forecast based on regression. Happily, there's a whole arsenal of techniques for dealing with it as part of a huge and fascinating discipline called time series analysis. The bad news is that it needs specialist knowledge to wield it. Worse still, even those who have it can and do still end up in trouble.

Take the salutary tale of Google Flu Trends (GFT), which created a stir through its supposed ability to give early warning of outbreaks of this killer disease. In a paper published in the journal *Nature* in 2009, data analysts at the eponymous tech company and experts from the US Centers for Disease Control (CDC) claimed to have spotted the 2007/08 flu season a week or two before CDC's own detection network.[2] They'd done it by trawling through years of data stored in Google's colossal historical archive, hunting correlations between outbreaks of flu and search terms typed into the company's search engine. Rather than try to guess what terms were most predictive, the team handed the job to computers, which tried out a staggering 450 million models. The best one used 45 search terms to produce an impressive correlation with future outbreaks of 0.97.

It was an impressive display of number-crunching, and for a while GFT seemed to herald a new era, in which huge data sets and computing power breathed new life into tired old techniques like regression and correlation. Yet their associated 'terms and conditions' were as potent as ever, and quickly asserted their authority. Barely had the GFT algorithm been taken out of its box than it fouled up, missing a 2009 flu outbreak completely, forcing its creators to put out a software patch. It didn't make much difference: GFT's predictions remained barely better than the CDC's traditional methods, and had a habit of overestimating the size of outbreaks. In 2014, a team of old-school data analysts published a scathing analysis of GFT's performance, highlighting its inadequacies. They included failure to deal with the well-known problem with time series, autocorrelation.[3] The following year, Google shut down the GFT website, and offered its data set to anyone who thought they could do better. It's entirely possible there is useful 'signal' buried in there somewhere; what's less clear is how best to extract it – and whether it's even worth the effort.

But there's one insight which cannot be gainsaid. Even before the launch of GFT, there were claims that colossal data sets meant it was no longer necessary to fret about 'terms and conditions' – or indeed even know what one was doing. Instead, the data could just be shovelled into computers, which would compare everything with everything else until they found the best possible correlations. No need for understanding, models or even hunches; as one wide-eyed commentator put it: 'With enough data, the numbers speak for themselves.'[4] The GFT debacle has shown that to be, as one celebrated data expert put it, 'Complete bollocks. Absolute nonsense.'[5] The fact is that this bright new field of 'Big Data' is subject to the same tiresome but crucial 'terms and conditions' as Small Data – along with extra pitfalls for good measure. Anyone thinking of picking up a digital shovel and digging into vast data sets should bear this in mind. If even Google's finest can end up with little more than fool's gold, just think what data mining could do for you (see box opposite).

None of this has deterred the cheerleaders for Big Data. With an evangelical belief in the miraculous power of methods such as regression, they have made it big news among big business. In 2014 a global survey found that around three-quarters of organisations will have invested in Big Data technology by 2016; the market is already worth around $125 billion.[6] The top priorities are to use the technology for 'enhancing the customer experience' and 'improving process efficiency'. Yet there are already signs of big trouble ahead. Industry insiders are warning that companies plan to mine pretty much everything and anything in their data archives – a sure-fire

Warning: data mining in progress

Data mining is a $100 billion global business, and everyone from multinationals to mom-and-pop outfits is scrambling to use it. So why are so many veterans of data analysis less than ecstatic about the Big Data revolution? Having spent their careers trying to extract insights from handfuls of data, surely they should be relishing getting their hands on truly colossal data sets. Yet decades of 'making do' with not very much have taught them some hard lessons that apply to all data sets, large or small. Take the problem of bias: a billion data-points from selective sources are potentially far more misleading than a tiny fraction obtained from a properly randomised sample (for example, just who are the people seeking flu remedies on Google, and why?).

Still, once you've got a clean, unbiased data set it's easy to create a forecasting model from it. Just use computer-based regression and correlation analysis to find statistically significant influences, then combine them to get a perfect fit to the data. On the contrary, that way lies disaster. When a data set is left to 'speak for itself' like this, it typically spouts nonsense. With no attempt to weed out implausible correlations, one ends up trusting 'statistical significance' to gauge relevance. Woefully inadequate at the best of times, it can prove catastrophic here. Pairing off just ten variables against each other while hunting for 'real' correlations carries a 90 per cent risk of finding at least one that's statistically significant purely by chance. Data mining often involves far greater numbers of variables. One way to cut the risk is to tighten the standard for significance. That helps, but then a very odd phenomenon emerges: the Jeffreys–Lindley Paradox. Long notorious among statisticians, this implies that the bigger the data set, the *less* effective significance tests are at spotting fluke findings. Another nasty surprise awaits those who think

▶

forecasting algorithms should ideally include as many variables as possible. While they'll give an impressive fit to archived data, such algorithms can fail badly once they go live. The problem lies in the so-called bias-variance dilemma. More variables give more accurate, less biased forecasts that fit old data well, but struggle with fresh input. As each variable has its own uncertainty, the fuzziness ('variance') of the forecast also increases. So a trade-off is needed: just enough variables to do a good job, but not so many that the forecasts are hopelessly vague.

All these challenges can be dealt with – if they're recognised from the outset. Contrary to what some might claim, when it comes to data mining, size isn't everything.

strategy for finding fool's gold. In the end, however, Big Data will live or die in the business world according to the time-honoured criterion: does it boost profits? That is far from guaranteed. One of the first big stories about Big Data centred on a $1 million prize offered by streaming movie service Netflix in 2006 to anyone who could data-mine a better way of predicting movie ratings. Three years later a team bagged the prize, but Netflix never went live with the algorithm. Despite meeting the required 10 per cent hike in performance, it was incredibly complex and the company baulked at paying for the IT upgrade needed to get so little benefit.[7] As data mining rolls out into the wider world, it will face similarly tough encounters with reality. Sales directors may not know of the perils of autocorrelation, but they do know when their data-mined sales forecasts are repeatedly way off.

To those keen to use the power of data mining, the concerns raised by old-school data analysts come across as reactionary and picky. After all, haven't scientists been using techniques like regression in research for decades, with no obvious problems? While scientists have indeed been heavy users of such techniques, the reliability of what they've found is much less certain. It would certainly be wrong to think scientists always wield the tools of data mining with care. A case in point is the salutary tale of Power Law Fever. During the 1980s, leading scientific journals began receiving papers claiming that phenomena from market movements to foraging by ants follow so-called power laws of the form

Something interesting = k × (something measurable)^N

The papers focused on finding the value of the power N, as this led to a host of interesting theories and ideas. To find out what it was, researchers used a simple trick allowing them to apply the method of linear regression to all kinds of data sets.[8] The resulting values of N spawned another wave of papers dedicated to explaining how and why these power laws existed. By the mid-1990s, one leading power law exponent felt emboldened to pen a popular-level book about it all with the modest title *How Nature Works*.[9] Even at the time, such claims raised eyebrows, but it took much longer for the qualms to turn to criticism. Quite why is an interesting question in the sociology of science, as it was clear from the outset that some researchers were doing extreme violence to several of the 'terms and conditions' of linear regression.[10] In their determination to find N, they risked reaching shockingly unreliable conclusions. Some of the most eye-catching were claims that power laws underpinned the behaviour of an astonishing variety of organisms. From the 1980s onwards, researchers claimed to have found that the foraging and hunting patterns of many creatures follow patterns known as Lévy flights. Mapped out, these look like random clusters of short excursions followed by rarer long ones – the relative proportions of each following a power law.

Various explanations were put forward, all essentially arguing that the mix of short and long hops was somehow 'optimal' for foraging. And it seemed it was being exploited by a a host of organisms from bees and albatrosses in the air, to ocean-going plankton, seals and even human fishing crews. But this 'evidence' had largely been based on linear regression – which can just go haywire when fed with data from such phenomena. In 2005, the mathematical ecologist Andrew Edwards, now at Fisheries and Oceans Canada,

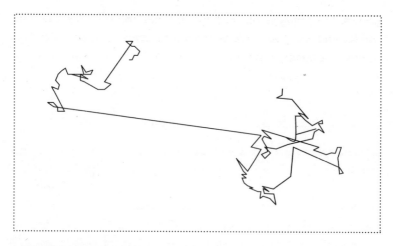

Is it a bird, is it a scribble? No, it's a Lévy flight – and a warning about pushing data too far

began investigating the basis of the claims, and re-analysed them using techniques better able to cope with the tricksy nature of power laws. He found that, out of seventeen published claims, not one of them stood up to scrutiny.[11] Since then, researchers have begun re-examining the whole subject using the more advanced methods and – in the case of the albatross at least – they've found that the original claims may have been right, albeit for the wrong reasons.[12]

That's good news for the ecologists, but leaves unanswered the question: just how much more regression-based baloney is still out there, as yet unrecognised? Unless someone decides to go back and check, we're unlikely ever to know. And with so much research and so many reputations now resting on regression-based results, it will be a brave researcher who decides to find out.

⬆ UPSHOT

All data sets contain patterns, but most are illusory. Finding the 'best-fit' line doesn't change that. Despite the hype, Big Data remains vulnerable to *GIGO – Garbage In, Garbage Out. Add in the oft-ignored 'terms and conditions' of data mining methods, and you've got a twenty-first-century technique for generating vintage nonsense.*

Playing the markets *isn't* rocket science

When people find that several trillion dollars they thought they'd stashed away has vanished, they tend to want answers. In 2007 the world's wage slaves had around $27 trillion in their pension funds. Most were counting on the money to give them a reasonable quality of life after decades of the nine-to-five. Many were already living off their modest nest eggs, counting on the stock markets to keep topping it up through growth and dividends. Then the financial crisis struck, the stock market collapsed and the value of the world's pension funds plunged by $3.5 trillion.[1]

In the search for culprits, the spotlight fell immediately on Wall Street, the City of London and investment banks everywhere. Then it focused on the Porsche-driving, bonus-grabbing denizens of these temples of greed, with their gelled hair and get-rich-quick schemes. But soon it turned to those who had dreamed up these schemes: the geeky 'rocket scientists' with their rimless spectacles and PhDs. While others have since had their time in the spotlight, it is these quantitative analysts – 'quants' – who have stayed there. They have been blamed for creating an arsenal of Weapons of Monetary Destruction that took the global economy to the brink of disaster. Many early accounts of the crisis focused on how the quants had been indulging in 'financial engineering', creating what they called 'derivatives' with bizarre names like credit default swaps and Bermuda swaptions. They'd also created so-called securities – obviously an in-joke – with alphabet soup names like ABS-CDOs.

But it was the internal workings of these WMDs that drew real gasps of horror. They were packed with hideously complex mathematical models that bore a striking resemblance to theoretical physics. The conclusion seemed clear: the global financial system had been captured by mad scientists.

This scary scenario has since been challenged on several fronts. First, derivatives are nothing new: the basic idea of a financial promise backed by a contract to deal with any default goes back millennia.[2] Secondly, far from being simply get-rich-quick schemes, they have long been seen as essential to commerce, bringing at least a modicum of confidence about an uncertain future. But the idea that investment banks were ever rammed with physicists dreaming up ever more crazy 'instruments' is also something of a myth. In reality, while there's no shortage of mathematically savvy people in finance, there's only ever been a relatively modest number of physicists. It's a crucial distinction, first because physicists have been among the most prominent critics of the use of complex financial techniques,[3] but also because they know the dirty little secret about their chosen discipline.

The truth about physics, and its central importance to understanding the financial crisis, is the focus of one of the most insightful analyses of the debacle, published in 2010. It bears the curious title 'Warning: Physics Envy may be hazardous to your wealth!',[4] and the credentials of its authors are no less intriguing: Andrew Lo, a distinguished professor of finance at the Massachusetts Institute of Technology Sloan School of Management, and Mark Mueller, an MIT physicist who had quit the field in the 1990s to become part of the very community charged with trashing the global economy: the POWs, or 'Physicists on Wall Street'. Together they examined the idea that the financial crisis was the result of the syndrome in the title of their study.

Despite its jokey name, Physics Envy is a genuine syndrome – and small wonder. Of all the sciences, none has achieved more success, credibility and kudos than physics. Its insights underpin the modern world and inform our notions of reality. Its greatest practitioners are bywords for genius, their grand theories hailed as among the greatest achievements of human intellect. Who wouldn't want a piece of that? In the aftermath of World War II, as physicists basked in global gratitude for helping to defeat evil, academics in other fields began wondering what the Way of Physics could do for them. Perhaps they, too, could identify fundamental laws and use them to model reality and thus shape the future to the benefit of humanity?

Among them was Paul Samuelson, a brilliant economics student who went up to the University of Chicago at 16 and at 22 had completed his PhD at Harvard. And not just any PhD: published in 1947 under the title *Foundations of Economic Analysis*, it became precisely that – and led to Samuelson becoming the first American to win the Economics Nobel. This was recognition of – as the prize citation put it in 1970 – his 'scientific work' and success in 'raising the level of analysis in economic science'. Samuelson was the catalyst for a radical shift in economics and finance, away from handwaving arguments and 'common-sense' reasoning and towards the principled and mathematical approach that had served physicists so well for so long. In truth, Samuelson was following in the footsteps of others who thought physics had much to teach economists. At the start of the twentieth century, a French mathematician named Louis Bachelier had applied probability theory to stock prices, finding evidence that they seemed to behave as if under the influence of random forces. Five years later, Einstein would develop a similar explanation for the jitters of microscopic particles – and used the results to infer the reality of atoms. Samuelson's own mentor Edwin Wilson was a scientific polymath and protégé of the brilliant American physicist Josiah Willard Gibbs.

During the 1950s and 1960s, economics and finance became ever more physics-like, their journals ever more impenetrable to those unfamiliar with the tools of physics, such as linear algebra and calculus. Yet as the twenty-something Samuelson saw, the similarities were dangerously misleading. As an initiate into the dirty little secret of physics, he knew that for all its apparent complexity and sophistication, physics succeeds because it focuses on problems that are essentially *simple*. This may seem risible to anyone who quit physics when lessons focused on parabolic trajectories of balls thrown off cliffs. Yet while the quadratic equations may seem abstruse, the fact they work at all is because the problem can be made *so* simple that equations can provide useful insight. Add in some realism – such as air resistance – and the maths rapidly becomes mind-bending.[5] Economics and finance don't share this level of complexity with physics – they are *far more* complex, replete with phenomena that simply defy the use of mathematics. Einstein's theoretical analysis of the behaviour of atoms was a fundamental breakthrough because its key features are forever true. Yet paradoxically that also made Einstein's task far simpler. Imagine if instead atoms sometimes took it upon themselves to rush in one direction, or to respond differently to identical forces. That would be a far harder, messier and less 'fundamental' problem to tackle. But it would also be one strikingly relevant to the behaviour of the 'atoms' of the stock market: human investors.

Samuelson saw the physics approach to economics and finance as helpful, but only up to a point. He understood that for all their grand theorising, physicists actually have it easy. Many of their greatest triumphs are based on the exploitation of symmetry – in essence, the ability of anything to be altered in some way, yet left unchanged. Being able to assume such constancy in space and time hugely simplifies theories of everything from subatomic particles to the entire universe.[6] Economists have nothing to match this: in their universe the only constant is change. In fact, as Lo and Mueller pointed out, the situation is even worse. As well as not being able to count on the constancy of the 'atoms' in their theories, economists cannot even tell whether or when their theories even apply.

The tragedy of economics in the post-war years was that the dazzling success of physics blinded so many to its dirty little secret. Too many economists came to regard the extensive use of maths by physicists as a sign of sophistication, rather than a symptom of simplification. Physicists are to be envied in that they get kudos for exploring worlds that will tolerate simplification to the point where they can apply maths – and still be left with something worth saying. They can tip away Nature's bathwater with impunity, as finding just a plastic duck is enough for them. Economists, in contrast, want to know whether babies are happier splashing around in pricier baths. They too can tip away the bathwater, but they risk also losing the baby, and are sure to end up with meaningless insights either way. Instead of envying physicists for the simplicity of what they do, economists envied their success at doing it. They did at least recognise the far greater uncertainty of problems in economics and finance. Forecasting, investing, designing derivatives all required that a view be taken on the uncertain future. In their quest to 'mathematise' their subject, economists thus headed to the theory of probability. But the most basic version was clearly not sufficient. Economics deal with situations far more complex than coin-tosses or dice-throws, where the probabilities are fixed and obvious. Financial markets are the result of multiple influences, all subject to uncertainty. As such, understanding them demanded that economists go up to the next level of probability theory, which captured the effect of multiple random influences. As we've seen, that led to heavy use of the Normal Distribution, whose elegance and power in dealing with messy real-life uncertainty had been recognised for over a century.

But as we've also seen, every technique for dealing with uncertainty comes with terms and conditions – some of which were clearly violated in the situations economists used them for. Evidence of these failings could be found in empirical data, but for years academics who pointed it out found their research rejected by leading economics and finance journals.[7] And there was an even more fundamental problem – one that simply could not be mathematised away. Looming over the uncertainties of economic phenomena and the models of those uncertainties was something far bigger: uncertainty about the models themselves. As Lo and Mueller point out, this puts economists in territory where even physicists walk in fear: where trusted models break down and have to be replaced. Physicists have faced such situations many times over the centuries. Galileo's laws of motion cracked when pushed too far, and gave way to Einstein's Special Theory of Relativity;

Newton's view of space, time and gravity became subsumed into Einstein's General Theory of Relativity; the view of atoms as tiny solar systems gave way to the probabilistic fuzz-balls of quantum mechanics. Physicists picked themselves up, learned the limits of the old models, and used them to select the best one for each job.

But economists can wake up tomorrow to find the equivalent of a suspension of the law of gravity. Yesterday, everything was fine; today Russia, say, defaults on its sovereign debt, causing some markets to crash like rocks experiencing an inverse square-*root* law of gravity. Meanwhile, others zoom as if gravity has been switched off. The standard models no longer work, and though they may kick back in some time, no one can say when.

In the face of such model uncertainty, fancy maths is powerless. There's only one thing that can prevent disaster: judicious use of the most complex device in the known universe – the human brain. The shiny toys of partial differential equations and Itô calculus have to give way to solid experience, judgement and decisiveness.

The financial crisis had many causes, political, regulatory and psychological among them. Yet they all have their roots in the same phenomenon: humans trying to deal with uncertainty. The much-reviled 'rocket scientists' dealt with it by using ever more complex models and hoping the devil was not in the details. Others dealt with it by trying to make so much money there would at least be no uncertainty about their own future. But they were all outnumbered by the line managers and CEOs, regulators and lawmakers who willingly fell under the spell of Physics Envy, and the belief that the tricks that reveal cosmic secrets must surely work in finance. Even now, it is unclear how many of them have finally woken up to the fact that dealing with the uncertainties of the financial world demands expertise far beyond the merely mathematical.[8]

Most physicists take pride in being part of the most successful of all scientific disciplines, while remaining aware of the limitations of their modus operandi. Perhaps more of them should join with the likes of Lo and Mueller in letting others in on the dirty little secret of physics – before it blows another dirty great hole in the global economy.

↑UPSHOT

The financial crisis was a multi-trillion-dollar demonstration of the dangers of Physics Envy. While sophisticated mathematics of the type wielded by physicists may be necessary in finance, it's certainly not sufficient. Physics can count on certainties, while finance involves not just a host of uncertainties, but also uncertainty about those uncertainties.

Beware geeks bearing models

I f the best and the brightest in finance cannot be trusted to keep our money safe, what can the rest of us do? The first lesson of the financial crisis seems clear enough: be deeply sceptical of anyone claiming to have tamed uncertainty. This is easier said than done, as such people often come with PhDs, models of byzantine complexity and even hard evidence of their success stretching back years. In physics, all this would indeed be impressive, as it would be evidence of a real advance of lasting value. But this is not physics, with its fundamental laws and universal constants. This is finance, where models can only ever be simulacrums of certainty. They may indeed work, but only while their terms and conditions hold, and no one knows how long that will be. It could be decades or days.

The temptation to ignore all this and drink the Kool-Aid is crystallised in the fortunes – in every sense – of hedge funds. These secretive institutions are famed for hiring the smartest geeks to devise 'hedging' strategies designed to give maximum returns for minimum risk. They are also renowned for their so-called 2-and-20 business models, whereby clients pay hedge funds 2 per cent of assets under management simply to be in line to benefit from the hedge fund's collective brilliance, plus 20 per cent of any profits that this brilliance actually delivers. And if the financial press is any guide, it's a price well worth paying: hedge funds routinely make headlines for their skill in spotting opportunities and dodging calamities. But of course the press is no guide at all: it focuses on exceptional performers who then 'regress to the mean' – and the record shows this mean performance is no better than that achieved by standard investment strategies once the hefty fees have been subtracted.[1] In short, to invest in the typical hedge fund is to buy expensive proof that even the most complex financial models are themselves subject to uncertainty. The real brilliance of hedge funds lies in their business model, which guarantees fee income for as long as they can convince investors to keep faith with the strategies – which can be longer than the strategies keep working.

Fortunately, investing in hedge funds is largely a game only rich investors can play. The rest of us are likely to end up with investments managed according to Nobel Prize-winning strategies. Unfortunately, this is not cause for celebration, as the strategies emerged from one of the most egregious attempts to reduce the complexities of finance to the simplicities of physics. At the University of Chicago in the early 1950s, a twenty-something economics student keen on physics set out to do for investment portfolios what Newton had done for moving bodies. The result would win Harry Markowitz a share of the 1990 Economics Nobel. Back in the 1950s, expert advice on investing was as simple as it was preposterous: find a top-performing stock and put all your money into it. Markowitz knew this made little sense, and so did most investors. They had twigged it made much more sense to have a 'diversified' mix of assets in a portfolio, to spread the risk of losing everything if one tanks. But anyone who sat down to create such a portfolio immediately ran into a problem: what should the mix be? Half in racy shares, half in boring-but-safe government bonds? Or is that too safe? What about an 80/20 split between shares and bonds ... or maybe 60/30/10, with the 10 per cent in cash that can be accessed in a trice? Markowitz realised that such questions fell within the ambit of a branch of applied maths called constrained optimisation. What he needed to do was find the optimum mix of assets that minimised risk, while still making a decent return.

The equations he wrote down became the foundation for what's now called Modern Portfolio Theory (MPT). And on the face of it, they achieve something miraculous. You feed them with historical data about the assets in your portfolio, and they reveal the optimal mix of assets you should hold. But despite its name, MPT isn't a theory; it's a model, and as such is replete with 'terms and conditions' and assumptions that range from questionable to flat wrong.

Take the concept of 'risk'. Most people might think minimising risk means minimising the chances of suffering a long and sustained loss. Yet in searching for a way to model risk mathematically, Markowitz picked the statistical concept of variance – a measure of the jitters of an asset's value about its average value. This seems oddly abstruse, but Markowitz stuck with it because it allowed him to exploit a neat theorem in probability that unlocked the whole optimisation problem. Put simply, the theorem provided the link between the total risk of a portfolio and the risk of each asset it contains, and how they're correlated to each other. Or at least, it did if you believed with Markowitz that variance is a good measure of 'risk'. If you did, you were duly rewarded with a mathematical description of the key features of investing: the risks and returns of assets, and even the way they moved with or against each other.

Markowitz's equations confirmed the common-sense idea that it makes sense to have a mix of assets, but they went farther, showing precisely what 'good' diversification looked like. It required assets with low or preferably negative correlations to each other. That also made sense, as when one fell in value, others would rise to make up the losses. The equations also contained a few surprises – such as the benefits of including riskier assets. If these were anti-correlated with the others, they could actually *reduce* the overall risk of the portfolio.

All that even an amateur investor needed to do to exploit the power of MPT was to examine the past performance of some assets and establish their returns, degree of correlation and their risk (as measured by the variance in their returns[2]). Plugged into Markowitz's equations, they would be magically converted into the percentage splits between the assets needed to create the optimal portfolio, diversified to minimise the risk while achieving a decent return.

But as countless investors have discovered over the years, apart from confirming the value of diversification, MPT raises more questions than it answers. Is variance really a good measure of 'risk'? After all, variance includes jitteriness above as well as below the mean value, and investors rarely worry about the former. Can't MPT give us a better measure of risk – such as the chance of the portfolio losing some percentage of its value? In theory, it can – if we assume the returns follow some probability distribution. But which one, and how can we tell when it no longer works?

Then there's the problem of the values of the returns, variance and correlations we feed into the equations for all the assets. If this were physics, we could just look them up in tables, as they'd be constants, like the masses of electrons and protons. But the only constant of assets is their constantly changing return and volatility. It's possible to work out average values – but over what timescales should they be calculated, and what if they're following distributions where the very notion of variance doesn't even make sense?

Correlations are another huge source of uncertainty; even rules of thumb such as bonds being anti-correlated with stocks hold true only until they don't.[3] Worse, the herd mentality of investors means that anti-correlated assets often move into synch precisely when their diversity is most needed, such as during a financial crisis.[4]

In the face of all these challenges, many investors have found it hard to trust the mathematics of MPT – among them Markowitz himself. Shortly after developing the theory, he was faced with setting up his own retirement account. He should have analysed performance records and calculated the optimal mix but found he couldn't face the prospect of being wrong – and simply put half his money into shares, and the other half in bonds.[5]

In the decades since its emergence, there have been many attempts to make MPT more sophisticated. The result has been a huge technical literature but little improvement beyond the core idea that diversification makes sense. In the end, no amount of mathematics can give MPT – or any strategy for investment – the reliability of physics. They will always remain models of uncertain phenomena whose validity is itself uncertain. In recent years, this has lent weight to the argument that there is simply no point trying to create portfolios and applying 'active management', buying, selling and tweaking the mix in an attempt to outperform the stock market. It's a belief backed by evidence that many apparently successful 'active' investors are – like hedge funds – nothing more than outliers whose performance regresses to the mean.[6] Even those who do beat the market typically fail to do so by a margin capable of justifying the fees charged.[7]

All this has led some of the smartest brains in finance to argue that the best strategy is the simplest. David Swensen, CIO of Yale University's $24 billion endowment fund, leading quant Paul Wilmott and investing legend Warren Buffett have all expressed enthusiasm for portfolios that simply mimic market performance using so-called 'index tracker' funds.[8] As the name suggests, these are set up to track the ebb and flow of market indices like the US S&P500, UK FTSE 100 or MSCI World Allcap using computers. Such 'passive' funds can never outperform their index, though this needs to be weighed against the fact that the S&P500 has averaged a respectable 8 per cent annual real-terms growth since 1985 – and active funds typically fail to do that, net of their fees for being allowed to try. Nor can passive funds free us of the need to diversify; several covering different areas are typically needed to dodge the worst volatility. But crucially, with so little human intervention, they do charge very low management fees – the one certain threat to portfolio performance.

The passive approach also addresses arguably the most important source of investment underperformance: ourselves. Many people regard investing as just a higher form of gambling. Pouring money into a handful of assets with zero diversification would certainly justify that view. But gambling is also notorious for how it affects the mind, its probabilistic nature triggering a host of potentially disastrous behaviours: taking too much risk when we win, chasing losses when we don't and persisting with half-baked strategies with no attempt to gauge success or failure. The inherent uncertainties of investing are known to affect the mind in similar ways. Studies suggest that most investors respond to its probabilistic nature by being either too confident or too complacent for their own good.[9] This leads to a plethora of wealth-destroying behaviours: attributing a few lucky breaks to genuine skill, backing expensive 'winners' which then regress to the mean, mistaking short-term 'noise' for long-term insight. Successful investors – like professional gamblers – have found ways of controlling these behaviours. These may well be beyond most of us, especially in times of crisis, in which case we may do best by doing as little as possible. The time-honoured way of doing that is via 'buy and hold' investing, which simply means deciding on our portfolio and then leaving well alone. There's plenty of evidence that many of us can – and do – do worse.[10] A recent study of US mutual fund investor records revealed that their tendency to buy booming stocks and sell sliding ones costs them dearly. Between 2000 and 2012 those who tried to spot winners and losers earned an average annual return of 3.6 per cent. In contrast, those who simply bought and held their portfolio earned 5.6 per cent.[11] A couple of per cent per year may not sound much, but if maintained over several decades it would compound into a 77 per cent boost to a portfolio's worth. Perhaps the most compelling evidence for the 'doing less is more' approach comes from that most celebrated of investors, Warren Buffett. In one of his famous Letters to Shareholders he revealed one of the cornerstones of his success in dealing with the risks and uncertainties of investing: 'Lethargy bordering on sloth'.[12]

So is buying and holding index trackers the path to successful investing? There is certainly evidence that it *can* work, but in the end it's still just a *model* for successful investing. And that means its approach to tackling uncertainty is itself subject to uncertainty. For example, in the fifteen years from the start of 1985 to the end of 1999, the S&P500 averaged a staggering annual growth rate of 15 per cent in real terms. Passive investors made out like bandits, seeing their portfolio values increase eightfold. Yet those who then signed up to the passive agenda have spent the last fifteen years struggling to average 2 per cent growth each year, and ended up with portfolios barely 30 per cent bigger. What had they done wrong? Nothing – apart from failing to foresee that the passive model was about to let them down, via two of the worst crashes of the last 100 years: the bursting of the Dotcom Bubble in 2000, and the financial crisis of 2007/08. During those times, the passive model demanded that believers simply sit and watch their portfolios collapse. Meanwhile, many active model veterans were able to draw on their experience, preserve value as the indexes fell, spot bargains – and bounced back big time.

This brings us full circle in our exploration of chance and uncertainty and how to deal with their myriad manifestations. The single most important rule is this: never lost sight of the fact that even if, through skill or luck, we find the Right Thing To Do, there is always a chance it will let us down. Our reluctance to accept this has led to endless misery, recrimination and guilt. Yet we should beat ourselves up only if we have failed to consider what we'd do if the Right Thing goes wrong. All we can ever do is give ourselves the best *chance* of success, accept this is always less than 100 per cent, and prepare for that eventuality.

In the end, we must all roll the dice and take our chances.

Acknowledgements

The depth, breadth and scope of the laws of probability are astounding. Every aspect of them, from their history and interpretation to their theoretical foundations and practical applications, could form the basis of a lifetime's work. Of all the disciplines I have wrestled with over 30 years as a science writer and academic, probability remains the one that continues to intrigue me and leaves me determined to learn more. I have also discovered that it has the same effect on those who study and use it professionally – creating a community of researchers and practitioners with an unusual mix of traits. They typically have brains the size of a planet, combined with beguiling modesty and willingness to help anyone hoping to understand the ways of randomness, risk and uncertainty. It has been a privilege to spend time in their company over the years, benefiting from their experience and insights. I especially want to thank Doug Altman, Iain Chalmers, Steven Cowley, Peter Donnelly, Frank Duckworth, Gerd Gigerenzer, the late Jack Good, John Haigh, Colin Howson, the late Dennis Lindley, David Lowe, Paul Parsons, Peter Rothwell, Stephen Senn, David Spiegelhalter and Henk Tijms.

This book would not exist without the initial suggestion of Ian Stewart, the constant enthusiasm of John Davey of Profile Books, and the love and support of Denise Best, my partner, muse and best friend.

As for the errors in this book, they are all my own work, and I would welcome having them pointed out. Experience has taught me that the probability of my making zero errors in matters of probability is itself zero.

Notes

Chapter 1

1. J. E. Kerrich, *An Experimental Introduction to the Theory of Probability*, E Munkgaard, Copenhagen, 1946.

2. J. Strzałko et al., 'Dynamics of coin tossing is predictable', *Physics Reports*, 469(2), 2008, pp. 59–92.

3. P. Diaconis et al., 'Dynamical bias in the coin toss', *SIAM Review*, 49(2), 2007, pp. 211–35.

Chapter 3

1. Surprisingly, it's not infinity; it's actually −1/12, arguably the single most amazing result in mathematics.

2. S. Stigler, 'Soft questions, hard answers: Jacob Bernoulli's probability in historical context', *Intl Stat. Rev.*, 82(1), 2014, pp. 1–16.

3. An analogy may help here. World-class archers have high confidence of getting close to the bull's-eye with just a few arrows. In contrast, beginners have low confidence of getting close to the bull's-eye with the same number of arrows. Give them enough, though, and even they can have high confidence of getting some arrows close to the bull's-eye. The question Bernoulli cast light on was: what's the relationship between the level of confidence, the closeness to the bull's-eye and the number of attempts?

4. Stigler, 'Soft questions, hard answers'.

5. Bernoulli had tried to simplify the calculations when using his theorem, but they were too crude. De Moivre found better approximations, and in the process invented an early version of the Central Limit Theorem, which we'll meet later.

Chapter 5

1. For any trait (e.g. birthday or star sign) where everyone has an equal chance of being in one of G groups (G = 365 for birthdays, 12 for star signs), you need a crowd of N people to give evens odds of at least one exact match, where N is 1.18 times the square root of G. For the theory for other coincidences, see R. Matthews and F. Stones, 'Coincidences: the truth is out there', *Teaching Statistics*, 20(1), 1998, pp. 17–19.

Chapter 6

1. M. Hanlon, 'Eggs-actly what ARE the chances of a double-yolker?', *Daily Mail Online*, 3 February 2010.

Chapter 8

1. J. A. Finegold et al., 'What proportion of symptomatic side-effect in patients taking statins are genuinely caused by the drug?', *Euro. J. Prev. Cardiol.*, 21(4), 2014, pp. 464–74.

2. R. Matthews, 'Medical progress depends on animal models – doesn't it?', *J. Roy. Soc. Med.*, 101(2), 2008, pp. 95–8.

Chapter 9

1. B. G. Malkiel, *Vanguard* study results cited in B. I. Murstein, 'Regression to the mean: one of the most neglected but important concepts in the Stock Market', *J. Behav. Fin.*, 4(4), 2003, pp. 234–7.

Chapter 10

1. D. A. Graham, 'Rumsfeld's knowns and unknowns: the intellectual history of a quip'. *The Atlantic* (online), 27 March 2013.

2. R. A. Fisher, *The Design of Experiments*, Oliver & Boyd, Edinburgh, 1935, p. 44.

3. I. Chalmers, 'Why the 1948 MRC trial of streptomycin used treatment allocation based on random numbers', *JLL Bulletin*: 'Commentaries on the history of treatment evaluation', 2010.

4. B. Djulbegovic et al., 'Treatment success in cancer', *Arch. Int. Med.*, 168, 2008, pp. 632–42.

5. J. Henrich, S. J. Heine and A. Norenzayan, 'The weirdest people in the world?', *Behav. & Brain Sci.*, 33(2), 2010, pp. 61–83.

6. P. M. Rothwell, 'Factors that can affect the external validity of randomised controlled trials', *PLOS Clin. Trials*, 1(1), 2006, p. e9.

7. U. Dirnagl and M. Lauritzen, 'Fighting publication bias', *J. Cereb. Blood Flow & Metab.*, 30, 2010, pp. 1263–4.

8. C. W. Jones and T. F. Platts-Mills, 'Understanding commonly encountered limitations in clinical research: an emergency medicine resident's perspective', *Annals Emerg. Med.*, 59(5), 2012, pp. 425–31.

9. S. Parker, 'The Oportunidades Program in Mexico', *Shanghai Poverty Conference*, 2003.

10. A. Petrosino et al., '"Scared Straight" and other juvenile awareness programs for preventing juvenile delinquency', *Cochrane Database of Systematic Reviews*, 4, 2013.

11. For example, the Behavioural Insights Team works with the UK Cabinet Office on 'Nudge Theory' approaches to implementing policy. Many of its successes stem from its extensive use of RCTs; see www.tinyurl.com/Organ-Donation-Strategy.

Chapter 11

1. The UK National Health Service 'Behind the Headlines' website does a great job of debunking these claims; see, e.g., www.tinyurl.com/SleepingPillsAlzheimers.

2. World Cancer Research Fund International, 'Diet, nutrition, physical activity and liver cancer', *Continuous Update Project* report, 2015.

3. J. N. Hirschhorn et al., 'A comprehensive review of genetic association studies', *Genetics in Medicine*, 4(2), 2002, pp. 45–61.

4. R. Sinha *et al.*, 'Meat intake and mortality: a prospective study of over half a million people', *Arch. Int. Med.*, 169(6), 2009, pp. 562–71; M. Nagao et al., 'Meat consumption in relation to mortality from cardiovascular disease among Japanese men and women', *Euro. J. Clin. Nutr.*, 66(6), 2012, pp. 687–93; S. Rohrmann et al., 'Meat consumption and mortality-results from the European Prospective Investigation into Cancer and Nutrition', *BMC Med.*, 11(1), 2013, p. 63.

5. S. S. Young and A. Karr, 'Deming, data and observational studies: a process out of control and needing fixing', *Significance*, September 2011, pp. 122–6.

6. M. Belson, B. Kingsley and A. Holmes, 'Risk factors for acute leukemia in children: a review', *Env. Health Persp.*, 2007, pp. 138–45.

7. A. B. Hill, 'The environment and disease: association or causation?', *Proc. Roy. Soc. Med.*, 58(5), 1965, pp. 295–300.

Chapter 12

1. K. de Bakker, A. Boonstra and H. Wortmann, 'Does risk management contribute to IT project success? A meta-analysis of empirical evidence', *Intl J. Proj. Mngt*, 28, 2010, pp. 493–503; D. Ramel, 'New analyst report rips Agile', *ADT Magazine*, 13 July 2012; R. Bacon and C. Hope, *Conundrum: Why every government gets things wrong and what we do about it*, Biteback, London, 2013.

2. One of the best known is the Bradley Effect, named after the eponymous Democratic nominee in the 1982 election for the governership of California. It has since been identified as playing a role in opinion poll debacles such as the UK general elections of 1992 and 2015. Ironically, the Bradley Effect was more likely due to simple sampling error: his defeat was within 1 per cent of the total, easily swamped by 'Don't Knows' – another source of error in conventional polls.

3. L. Hong and S. E. Page, 'Groups of diverse problem solvers can outperform groups of high-ability problem solvers', *PNAS*, 101, 2004, pp. 16385–9.

4. C. P. Davis-Stober et al., 'When is a crowd wise?', *Decision*, 1(2), 2014, pp. 79–101.

5. A. B. Kao and I. D. Couzin, 'Decision accuracy in complex environments is often maximized by small group sizes', *Proc. Roy. Soc. B*, 281(1784), 2014, 20133305.

6. S. M. Herzog and R. Hertwig, 'Think twice and then: combining or choosing in dialectical bootstrapping?', *J. Exp. Psychol.: Learning, Memory, and Cognition*, 40(1), 2014, pp. 218–33.

Chapter 14

1. More on the theory of casino games and a host of other aspects of probability can be found in my favourite text on the subject: *Taking Chances* by John Haigh (Oxford University Press, 2003).

Chapter 15

1. J. Rosecrance, 'Adapting to failure: the case of horse race gamblers', *J. Gambling Behav.*, 2(2), 1986, pp. 81–94.

2. P. Veitch, *Enemy Number One*, Racing Post Books, Newbury, 2010.

Chapter 17

1. Let the chances of the rumour proving true be P; then the chances of the rumour proving false are $1 - P$ (as one of these two outcomes *must* be true, their chances must add up to 1). The expected consequences of staying put are then $-10P + 7(1 - P)$, while those for moving are $2P + (1 - P)$. Equating these two gives us the probability above which moving leads to more positive consequences. We then find that it makes sense to move if the chances P of the rumours being true exceed 1/3.

Chapter 18

1. Alice Thomson is a pseudonym for a real person contacted by the author in January 2015.

2. G. Gigerenzer, in *Reckoning with Risk*, Allen Lane, London, 2002, pp. 42–5.

3. K. Moisse, 'Man takes pregnancy test as joke, finds testicular tumor', *ABC News online*, 6 November 2012.

4. This is a simple consequence of Bayes's Theorem, described in Chapter 20.

Chapter 19

1. R. Matthews, 'Decision-theoretic limits on earthquake prediction', *Geophys. J. Int.*, 131(3), 1997, pp. 526–9.

2. R. Matthews, 'Base-rate errors and rain forecasts', *Nature*, 382, 1996, p. 766.

Chapter 20

1. This is based on the article 'A speck in the sea' by Paul Tough, *New York Times Magazine*, 2 January 2014.

2. For a popular-level account of Bayes's Theorem and its history and applications, see S. B. McGrayne, *The Theory That Would Not Die*, Yale University Press, 2011.

3. The original can be found at www.tinyurl.com/Bayes-Essay.

4. The formulas come from the so-called binomial distribution.

5. I focus throughout the book on the simplest form of the theorem, involving a straight dichotomy between one hypothesis and all the alternatives. It should be stressed, however, that Bayes's theorem can cope with far more complex cases.

6. For a carefully argued analysis of Bayes's struggle with the 'Problem of Priors', and the misconceptions that followed, see S. M. Stigler, 'Thomas Bayes's bayesian inference', *Journal of the Royal Statistical Society. Series A (General)*, 1982, pp. 250–58.

7. Contrary to what even many advocates of Bayesian reasoning think, however, the same evidence can drive the two camps in a controversy farther apart. See R. Matthews, 'Why do people believe weird things?', *Significance*, December 2005, pp. 182–4.

Chapter 21

1. I. J. Good, 'Studies in the history of probability and statistics. XXXVII: AM Turing's statistical work in World War II', *Biometrika*, 1979, pp. 393–6.

2. S. Zabell, 'Commentary on Alan M. Turing: the applications of probability to cryptography', Cryptologia, 36(3), 2012, pp. 191–214.

3. Y. Suhov and M. Kelbert, *Probability and Statistics by Example*, vol. 2: *Markov Chains: A Primer in Random Processes and Their Applications*, Cambridge University Press, Cambridge, 2008, p. 433.

4. This follows from applying logarithms to the original. The resulting formula does not appear explicitly in Turing's report, but 'log-transforming' is a key part of its arguments.

5. D. A. Berry, 'Bayesian clinical trials', *Nat. Rev. Drug Discov.*, 5(1), 2006, pp. 27–36.

6. M. Dembo et al., 'Bayesian analysis of a morphological supermatrix sheds light on controversial fossil hominin relationships', *Proc. R. Soc. B.*, 282(1812), 2015, 20150943.

7. R. Trotta, 'Bayes in the sky: Bayesian inference and model selection in cosmology', *Contemp. Physics*, 49(2), 2008, pp. 71–104.

Chapter 22

1. R. Matthews, 'The interrogator's fallacy', *Bull. Inst. Math. Apps*, 31(1), 1995, pp. 3–5.

2. S. Connor, 'The science that changed a minister's mind', *New Scientist*, 29 January 1987, p. 24.

Chapter 23

1. H. Jeffries, *Theory of Probability*, 1939, pp. 388–9; W. Edwards, H. Lindman and L. J. Savage, 'Bayesian statistical inference for psychological research', *Psychol. Rev.*, 70(3), 1963, pp. 193–242; J. Berger and T. Sellke, 'Testing a point null hypothesis: the irreconcilability of P-values and evidence', *JASA*, 82(397), 1987, pp. 112–22; R. Matthews, 'Why should clinicians care about Bayesian methods?', *J. Stat. Plan. Infer.*, 94(1), 2001, pp. 43–58; 'Flukes and flaws', *Prospect* magazine, November 1998.

2. See P. R. Band, N. D. Le, R. Fang and M. Deschamps, 'Carcinogenic and endocrine disrupting effects of cigarette smoke and risk of breast cancer', *Lancet*, 360(9339), 2002, pp. 1044–9, contradicted the following month by Collaborative Group on Hormonal Factors in Breast Cancer, 'Alcohol, tobacco and breast cancer', *B. J. Canc.*, 87(11), 2002, pp. 1234–45.

3. For an entertaining demonstration, see G. D. Smith and E. Shah, 'Data dredging, bias, or confounding: they can all get you into the *BMJ* and the Friday papers', *BMJ*, 325(7378), 2002, p. 1437.

4. G. Taubes, 'Epidemiology faces its limits', *Science*, 269(5221), 1995, pp. 164–9.

5. J. P. A. Ioannidis, 'Why most published research findings are false', *PLOS Medicine*, 2(8), 2005, p. e124.

6. J. P. A. Ioannidis, 'Contradicted and initially stronger effects in highly cited clinical research', *JAMA*, 294(2), 2005, pp. 218–28; R. A. Klein et al., 'Investigating variation in replicability: a "many labs" replication project', *Social Psychology*, 45(3), 2014, pp. 142–52; M. Baker, 'First results from psychology's largest reproducibility test', *Nature* online news, 30 April 2015.

7. 2014 Global R&D Funding Forecast (Batelle.org, December 2013).

8. R. A. Purdy and S. Kirby, 'Headaches and brain tumors', *Neurol. Clin.*, 22(1), 2004, pp. 39–53.

9. J. Aldrich, 'R A Fisher on Bayes and Bayes' Theorem', *Bayesian Analysis*, 3(1), 2008, pp. 161–70.

10. R. A. Fisher, 'The statistical method in psychical research', *Proc. Soc. Psych. Res.*, 39, 1929, pp. 189–92; Fisher explicitly describes the

arbitrary nature of the p-value standard for significance, and warns of the dangers of misinterpretation.

11. F. Yates, 'The influence of statistical methods for research workers on the development of the science of statistics', *JASA*, 46(253), 1951, pp. 19–34.

12. F. Fidler et al., 'Editors can lead researchers to confidence intervals, but can't make them think statistical reform lessons from medicine', *Psych. Sci.*, 15(2), 2004, pp. 119–26.

13. S. T. Ziliak and D. N. McCloskey, *The Cult of Statistical Significance: How the Standard Error Costs Us Jobs, Justice and Lives*, University of Michigan Press, Ann Arbor, 2008, ch. 7.

14. F. L. Schmidt and J. E. Hunter, 'Eight common but false objections to the discontinuation of significance testing in the analysis of research data', in L. L. Harlow et al. (eds), *What If There Were No Significance Tests?*, Psychology Press, Oxford, 1997, pp. 37–64.

15. When the author first began reporting on this issue in the 1990s, he was told by several learned bodies, including the Royal Statistical Society and the British Psychological Society, that clear policy statements on p-values were too provocative for their membership and journals.

16. D. M. Windish, S. J. Huot and M. L. Green, 'Medicine residents' understanding of the biostatistics and results in the medical literature', *JAMA*, 298(9), 2007, pp. 1010–22.

Chapter 24

1. J. Maddox, 'CERN comes out again on top', *Nature*, 310(97), 12 July 1984.

2. J. W. Moffat, *Cracking the Particle Code of the Universe*, Oxford University Press, Oxford, 2014, p. 113.

3. Sigma values are a measure of the degree of separation between the results obtained, and what would be expected if they were nothing but flukes. Thus, unlike p-values, the *bigger* the sigma value, the bigger the separation between the results and mere flukes. They're also highly non-linear measures of 'significance' with a jump of sigma from 2 to 4 corresponding to a 700-fold increase in 'significance'. We'll encounter them again in our treatment of the financial crisis.

4. D. Mackenzie, 'Vital statistics', *New Scientist*, 26 June 2004, pp. 36–41.

5. See, e.g., R. Matthews, 'Why should clinicians care about Bayesian methods?', *JSPI*, 94(1), 2001, pp. 43–58.

6. The results quoted are based on the theory in J. Berger and T. Sellke, 'Testing a point null hypothesis: the irreconcilability of P-values and evidence', *JASA*, 82(397), 1987, pp. 112–22 (especially section 3.5); given the distributional assumptions and lower bounds involved, the figures are indicative only.

Chapter 25

1. W. W. Rozeboom, 'Good science is abductive, not hypothetico-deductive', in L. L. Harlow et al. (eds), *What If There Were No Significance Tests?*, Psychology Press, Oxford, 1997, pp. 335–92.

2. Put simply, the problem lies in the fact that many research questions involve ranges ('distributions') of prior probabilities and also of alternative explanations of the data. In simple cases, one can use 'conjugate densities' giving formulas into which to plug data and prior insights, but many real-life applications demand computer-intensive techniques.

3. S. Connor, 'Glaxo chief: Our drugs do not work on most patients', *Independent*, 8 December 2003, p. 1.

4. S. J. Pocock and D. J. Spiegelhalter, 'Domiciliary thrombolysis by general practitioners', *BMJ*, 305(6860), 1992, p. 1015.

5. By itself, the 95 per cent CI means that if we took a large random sample (in this case, of participants in the trial) drawn from the same population (in this case, all suitable patients), we could be confident the resulting CI would cover the population value of whatever we're interested in – say, an odds ratio for risk of death – 95 per cent of the time (presuming, of course, all non-random sources of error, such as bias, have been eliminated). Thus the 'confidence' relates to the reliability of the *statistical technique*, not the reliability of the *finding*. Bayes shows we can take the former as a measure of the latter only if we're in a state of utter prior ignorance of what that finding might be – which is rarely the case. After decades of research, we usually do have prior insights to draw on, and Bayes then gives us a 95 per cent *credible* interval, where the credibility really does relate to the finding

(with the usual caveats of the trial being free from other sources of error).

6. L. J. Morrison et al., 'Mortality and prehospital thrombolysis for acute myocardial infarction: a meta-analysis', *JAMA*, 283(20), 2000, pp. 2686–92.

Chapter 26

1. S. Kühn and J. Gallinat, 'Brain structure and functional connectivity associated with pornography consumption', *JAMA Psychiatry*, 71(7), 2014, pp. 827–34.

2. J. A. Tabak and V. Zayas, 'The roles of featural and configural face processing in snap judgments of sexual orientation', *PLOS One*, 7(5), 2012, e36671.

3. I. Chalmers and R. Matthews, 'What are the implications of optimism bias in clinical research?', *The Lancet*, 367(9509), 2006, pp. 449–50. For the challenges of 'prior elicitation' in clinical trials, see D. J. Spiegelhalter, K. R. Abrams and J. P. Myles, *Bayesian Approaches to Clinical Trials and Health-care Evaluation*, Wiley, Chichester, 2004, pp. 147–8. Humans in general seem to be biased towards a rosy view of future events; see, e.g., T. Sharot, 'The optimism bias', *Current Biology*, 21(23), 2011, pp. R941–R945.

4. R. Matthews, 'Methods for assessing the credibility of clinical trial outcomes', *Drug Inf. Ass. J.*, 35(4), 2001, pp. 1469–78, www.tinyurl.com/credibility-prior; an online calculator is available here: http://statpages.org/bayecred.html.

5. H. Gardener et al., 'Diet soft drink consumption is associated with an increased risk of vascular events in the Northern Manhattan Study', *J. Gen. Int. Med.*, 27(9), 2012, pp. 1120–26.

6. The work of Ramsey and De Finetti in the 1920s and Cox and Jaynes in the 1960s all pointed to the ineluctability of the probability calculus for capturing belief; see C. Howson and P. Urbach, *Scientific Reasoning: The Bayesian Approach*, Open Court, Chicago, IL, 1993, ch. 5.

7. K. H. Knuth and J. Skilling, 'Foundations of inference', *Axioms*, 1(1), 2012, pp. 38–73.

Chapter 27

1. R. J. Gillings, 'The so-called Euler–Diderot incident', *Am. Math Monthly*, 61(2), 1954, pp. 77–80.

2. E. O'Boyle and H. Aguinis, 'The best and the rest: revisiting the norm of normality of individual performance', *Personnel Psych.*, 65, 2012, pp. 79–119; J. Bersin, 'The myth of the Bell Curve: look for the hyper-performers', *Forbes online*, 19 February 2014.

3. For events whose probability in a single trial is p (= 0.5 for a coin-toss), the chances of getting S successes in any order during x attempts are given by the binomial distribution: $[S!/(S-x)!x!]p^x(1-p)^{S-x}$, where '!' means factorial, which can be found on any scientific calculator. So the chances of getting exactly five heads from ten tosses is $[10!/5!5!](0.5)^5(1-0.5)^{10-5} = 0.246$. Factorials and powers become very tedious to work with for large S.

4. Strictly speaking, Laplace's 'classical' version of the theorem also puts constraints on the behaviour of these independent random influences. Mathematicians have since proved that the theorem still holds – and the Bell Curve eventually emerges – even when the random influences don't behave identically. Yet even under these so-called Lindeberg–Feller conditions, the influences must be independent and incapable of behaving too wildly – which are still major caveats.

5. Such arguments are often questionable, however: see A. Lyon, 'Why are Normal Distributions normal?', *B. J. Phil. Sci.*, 65(3), 2014, pp. 621–49.

6. S. Stigler, *Statistics on the Table*, Harvard University Press, Cambridge, MA, 2002, p. 53.

7. Ibid., p. 412.

8. Quoted in H. Jeffreys, 'The Law of Error in the Greenwich variation of latitude observations', *Mon. Not. RAS*, 99(9), 1939, p. 703.

Chapter 28

1. K. Dowd et al., 'How unlucky is 25-sigma?', ArXiv.org preprint: arXiv:1103.5672, 2011.

2. K. Pearson, 'Notes on the history of correlation', *Biometrika*, 13, 1920, pp. 25–45.

3. This is based on real-life data from the US census of 1999, reported and analysed by M. F. Schilling, A. E. Watkins and W. Watkins, 'Is human height bimodal?', *Am. Stat.*, 56(3), 2002, pp. 223–9.

4. As Schilling et al. (ibid.) show, if the difference between the means of several Bell Curves exceeds a certain multiple of the sum of the standard deviations, the combined Bell Curve will look distinctly dented. Adding more Bell Curves also tends to skew the whole shape of the composite curve, spoiling its symmetry.

5. See, e.g., R. W. Fogel et al., 'Secular changes in American and British stature and nutrition', *J. Interdis. Hist.*, 14(2), 1983, pp. 445–81.

6. See, e.g., B. Mandelbrot, *The Misbehaviour of Markets*, Profile, London, 2005, which proved prescient following the financial crisis. For an account of the aftermath, see A. G. Haldane and B. Nelson, 'Tails of the unexpected', in proceedings of *The Credit Crisis Five Years On: Unpacking the Crisis*, University of Edinburgh Business School, 8/9 June 2012.

7. P. Wilmott, 'The use, misuse and abuse of mathematics in finance', *Phil. Trans. Roy. Soc.*, Series A, 358(1765), 2000, pp. 63–73.

8. They include the so-called Value at Risk (VaR) method, developed by financial engineers in the late 1980s, and now part of the so-called Basel III international banking risk standards produced following the financial crisis. VaR involves financial institutions estimating the chances of making specific losses over a specific time frame. Such estimates are often based on historical data and simulations, which carry obvious risks. They have been soundly attacked by Nassim Taleb, author of *The Black Swan* (Penguin, London, 2008); see, e.g., www.fooledbyrandomness.com/jorion.html.

9. JPMorgan Chase *Annual Report*, April 2014, p. 31.

Chapter 29

1. This is true for any symmetric distribution – presuming, that is, that the average exists, which – as we'll see with the Cauchy Distribution – it may not.

2. D. Veale et al., 'Am I normal? A systematic review and construction of nomograms for flaccid and erect penis length and circumference in up to 15,521 men', *BJU Intl.*, 115(6), 2015, pp. 978–86.

3. O. Svenson, 'Are we all less risky and more skillful than our fellow drivers?', *Acta Psychol.*, 47(2), 1981, pp. 143–8.

4. S. Powell, 'RAC Foundation says young drivers more likely to crash', *BBC Newsbeat*, 27 May 2014.

5. This is reflected in the use of logarithms. Laplace's Central Limit Theorem shows that we get a standard Normal curve as a result of independent random influences adding together. Using logarithms retains this additive property for influences which really act multiplicatively.

6. E. Limpert, W. A. Stahel and M. Abbt, 'Log-normal distributions across the sciences: keys and clues', *BioScience*, 51(5), 2001, pp. 341–52.

7. See L. T. DeCarlo, 'On the meaning and use of kurtosis', *Psych. Meth.*, 2(3), 1997, pp. 292–307.

8. Some advanced texts point out that one can inadvertently create a Cauchy Distribution by forming the ratio of two Normally distributed variables where the denominator passes through zero. This can cause unrecognised havoc even in calculating such basic characteristics as the ratio's mean and standard deviation, let alone 'significance testing'.

9. Using the theory of the Bell Curve, it's possible to show that a 25-sigma event has a staggeringly low probability of 1 in 10^{137}, that is 1 followed by 137 zeros. According to the Cauchy Distribution, however, the chances are 1 in 77, in other words around 10^{135} times more likely than the Bell Curve calculation suggests. One must never forget that incredibly rare events can and do happen all the time; the chances of your having another 24 hours precisely like those you've just had are far smaller than 10^{137}. But then, no sane person tries to develop a theory able to predict such things; in finance, they do.

10. E. F. Fama, 'The behavior of stock-market prices', *J. Business*, 38(1), 1965, pp. 34–105.

11. Named after Paul Lévy (1886–1971), they are also sometimes called stable Paretian – or even just 'stable' – distributions. Fama came to use them after encountering the work of Benoit Mandelbrot.

12. Their behaviour can be tuned using four 'knobs' – parameters, in the jargon – which determine the location of their peak, their squatness, skewedness and – crucially – the 'fatness' of their tails. The latter is dictated by a number between zero and 2. When it's exactly 2, the

result is the Bell Curve, but for lower values the distributions have an infinite variance. When it hits exactly 1 it becomes the Cauchy curve, lacking both a mean and variance. Values below 1 give insane results.

13. For a fairly non-technical account with many real-life examples and insights, see M. E. J. Newman, 'Power laws, Pareto distributions and Zipf's law', *Contemp. Physics*, 46(5), 2005, pp. 323–51.

14. The threat posed by power laws to the reliability of business research is examined in G. C. Crawford, W. McKelvey and B. Lichtenstein, 'The empirical reality of entrepreneurship: how power law distributed outcomes call for new theory and method', *J. Bus. Vent. Insight*, 1(2), 2014, pp. 3–7.

Chapter 30

1. R. A. Fisher and L. Tippett, 'Limiting forms of the frequency distribution of the largest or smallest member of a sample', *Math. Proc. Camb. Phil. Soc.*, 24(2), 1928, pp. 180–90.

2. Such rules of thumb emerge naturally from power law distributions. A power law distribution of the form $p(x) = Cx^{-a}$ leads to a proportion X of the total amount of a quantity (say, the world's wealth) being tied to a percentage P of the total population, where $X = P^K$ and $K = (a - 2)/(a - 1)$. So, for example, $a = 2.2$ gives the famous trope of 'almost 80% of wealth is accounted for by just 20% of the total population'.

3. M. Moscadelli, 'The modelling of operational risk: experience with the analysis of the data collected by the Basel Committee', *Temi di discussione* (Economic working papers), 517, Bank of Italy Economic Research Department, 2004.

4. K. Aas, 'The role of extreme value theory in modelling financial risk', Lecture, NTNU, Trondheim, 2008.

5. M. Tsai and L. Chen, 'The calculation of capital requirement using Extreme Value Theory', *Economic Modelling*, 28(1), 2011, pp. 390–95.

6. K. Aarssen and L. de Haan, 'On the maximal life span of humans', *Math. Pop. Studies*, 4(4), 1994, pp. 259–81.

7. In N trials of a random event of probability P, the length of the longest continuous streak is L, and it satisfies the equation $N(1 - P)P^L = 1$.

See M. F. Schilling, 'The surprising predictability of long runs', *Math. Mag.*, 85, 2012, pp. 141–9.

Chapter 31

1. Contrary to widespread belief, correlation coefficients say nothing about how big a change is produced in one variable by changes in another. Nor is correlation only measurable for simple, linear relationships: the Spearman correlation can cope with so-called monotonic non-linear relationships, and even non-Normality.

2. For data sets with at least ten pairs of data, any correlation level whose absolute magnitude exceeds 0.62 will be 'statistically significant' to better than the usual $p = 0.05$ standard. Most of Vigen's headline-grabbing correlations pass this standard with ease – underlining yet again the inadequacies of the concept of 'statistical significance' for weeding out nonsense.

3. As if all this wasn't bad enough, the most widely used method of determining correlations has the assumption of Bell Curve behaviour built into its very core.

4. The weird notion of storks delivering babies appears in 'The Storks', a short story published in 1838 by Hans Christian Andersen, but the mythology appears to be far older. It's since been 'confirmed' using correlation analysis by various investigators, including the author: R. Matthews, 'Storks deliver babies (p = 0.008)', *Teaching Statistics*, 22(2), 2000, pp. 36–8, who uses it to highlight the inadequacies of *p*-values; see also T. Höfer and H. Przyrembel, 'New evidence for the theory of the stork'. *Paed. & Peri. Epid.*, 18(1), 2004, pp. 88–92.

5. M. H. Meier et al., 'Persistent cannabis users show neuropsychological decline from childhood to midlife', *PNAS*, 109(40), 2012, pp. E2657–E2664.

6. O. Rogeberg, 'Correlations between cannabis use and IQ change in the Dunedin cohort are consistent with confounding from socioeconomic status', *PNAS*, 110(11), 2013, pp. 4251–4.

7. There is, for example, evidence that the health risks of passive smoking may be lower than often claimed; see J. E. Enstrom, G. C. Kabat and G. Davey Smith, 'Environmental tobacco smoke and tobacco related mortality in a prospective study of Californians, 1960–98', *BMJ*, 326(7398), 2003, pp. 1057–67. This is not an academic point: if the risk from this common confounder has been

overestimated, it could lead to other sources of repiratory and heart disease risk being overlooked.

8. D. Freedman, R. Pisani and R. Purves, *Statistics*, 3rd edn, W. W. Norton, New York, 1998, p. 149. The phenomenon of changing variance rejoices in the name *heteroskedasticity* (from the Greek words for 'different' and 'scatter').

9. Backing for Pearson's concerns is given in W. Dunlap, J. Dietz and J. M. Cortina, 'The spurious correlation of ratios that have common variables: a Monte Carlo examination of Pearson's formula', *J. Gen. Psych.*, 124(2), 1997, pp. 182–93. For a review of the problem of ratio-based correlations in business, see R. M. Wiseman, 'On the use and misuse of ratios in strategic management research', in D. D. Bergh and D. J. Ketchen (eds), *Research Methodology in Strategy and Management*, vol. 5, Emerald Group Publishing, Bingley, 2008, pp. 75–110.

10. Such seasonal variations in temperature are principally the result of the tilt of the Earth's axis to its orbit around the sun. It's worth stressing that there *are* techniques for dealing with non-linear correlations, but not everyone who needs them knows about them – or uses them.

Chapter 32

1. Various definitions of 'best' are possible, but linear regression is based on the so-called Principle of Least Squares suggested by Gauss, which has some elegant properties. The basic idea is to make the least possible error in estimating one variable using another.

2. J. Ginsberg et al., 'Detecting influenza epidemics using search engine query data', *Nature*, 457, 2009, pp. 1012–14.

3. D. Lazer et al., 'The parable of Google Flu: traps in big data analysis', *Science*, 343, 2014, pp. 1203–5.

4. C. Anderson, 'The end of theory: the data deluge makes the scientific method obsolete', *Wired*, 23 June 2008.

5. The eminent British statistician Professor Sir David Spiegelhalter, quoted in T. Harford, 'Big data: are we making a big mistake?', *Financial Times*, 28 March 2014.

6. Survey by Gartner, www.gartner.com/newsroom/id/2848718, 17 September 2014; market value from *Forbes* report, '6 predictions for

the $125 billion Big Data analytics market in 2015', published online, 11 December 2014.

7. S. Finlay, *Predictive Analytics, Data Mining and Big Data*, Palgrave Macmillan, London, 2014, p. 131.

8. If the data pairs (x, y) follow a 'power law' relationship such as $y = ax^n$, then $\log(y) = \log(a) + n\log(x)$, which is the formula for a straight line with intercept $\log(a)$ and slope of n. Linear regression applied to the data pairs then gives the 'best' estimates for $\log(a)$ and n – this latter being the sought-for power.

9. P. Bak, *How Nature Works: The science of self-organized criticality*, Springer, New York, 1996.

10. For a comprehensive review of both the theoretical and empirical problems, see A. Clauset, C. R. Shalizi and M. E. J. Newman, 'Power-law distributions in empirical data', *SIAM Review*, 51(4), 2009, pp. 661–703. As with correlation, there *are* ways of relaxing some of the 'terms and conditions' of textbook linear regression, notably 'non-parametric' methods which will work without knowing the underlying distributions involved. But these can still stuggle with the wild behaviour of power law distributions.

11. A. M. Edwards, 'Overturning conclusions of Lévy flight movement patterns by fishing boats and foraging animals', *Ecology*, 92(6), 2011, pp. 1247–57.

12. N. E. Humphries et al., 'Foraging success of biological Lévy flights recorded in situ', *PNAS*, 109(19), 2012, pp. 7169–74.

Chapter 33

1. B. Keeley and P. Love, 'Pensions and the crisis', in *From Crisis to Recovery: The Causes, Course and Consequences of the Great Recession*, OECD Publishing, Paris, 2010.

2. A derivative contract for a Mesopotamian merchant has been dated to 1809 BCE; see E. J. Weber, 'A short history of derivative security markets', Discussion Paper 08.10, University of Western Australia Business School, 2008.

3. Eminent examples include Emanuel Derman, Paul Wilmott and Riccardo Rebonato. Derman is a former particle physicist at Columbia University and author of *Models. Behaving. Badly* (Simon & Schuster, New York, 2011). Wilmott is co-author with Derman of the *Financial Modeller's Manifesto*, and has a PhD in fluid dynamics

from Oxford University. Rebonato is author of the prescient *Plight of the Fortune Tellers* (Princeton University Press, 2007) and has a PhD in condensed matter physics.

4. A. W. Lo and M. T. Mueller, 'Warning: Physics Envy may be hazardous to your wealth!', *J. Invest. Mngt*, 8(2), 2010, pp. 13–63.

5. As air resistance varies with the speed of the projectile, which thus changes in response, it takes advanced calculus to work out the trajectory. Add in a moving target and the Earth's rotation and you've got ballistics – the focus of research by leading physicists in World War II.

6. A simple example of the use of symmetry is the description of a square piece of paper; if rotated through 90 degrees, it looks identical – it's been 'changed without being changed'. More subtle symmetries have subtle links to those other powerful principles of physics: conservation laws, the link being manifested by an astonishing mathematical result known as Noether's Theorem.

7. Lo and Mueller, 'Warning: Physics Envy may be hazardous to your wealth!', section 2.3.

8. For a review of what these skills are and how they might be implemented in finance, see ibid.

Chapter 34

1. A. W. Lo et al., 'Hedge funds: a dynamic industry in transition', *Ann. Rev. Fin. Econ.*, 7, 2015.

2. Another oft-used metric is the so-called *volatility* of an asset, given by the square root of the variance, known in statistics as the standard deviation.

3. For example, the correlation between the US S&P500 market index and long-term US Treasury bonds has switched signs 29 times from 1927 to 2012, and has ranged from –0.93 to +0.84. See N. Johnson et al., 'The stock–bond correlation', PIMCO Quantitative Research Report, November 2013.

4. See, e.g., N. Waki, 'Diversification failed this year', *New York Times Business*, 7 November 2008; S. Stovall, 'Diversification: a failure of fact or expectation?', *Am. Ass. Indiv. Inv. J.*, March 2010.

5. Quoted in J. Zweig, *Your Money and Your Brain: How the new science of neuroeconomics can help make you rich*, Simon & Schuster, New York, 2007, p. 4.

6. R. Ferri, 'Coin flipping outdoes active fund managers', *Forbes*, 13 January 2014.

7. Research by the UK Department for Communities and Local Government, cited in M. Johnson, 'We don't need 80% of active management', *Financial Times*, 11 May 2014.

8. Quoted on the *Monevator* blog, 'The surprising investment experts who use index funds', 10 February 2015.

9. K. H. Baker and V. Ricciardi, 'How biases affect investor behaviour', *Euro. Fin. Rev.*, 28 February 2014.

10. J. Kimelman, 'The virtues of inactive investing', *Barron's*, 10 September 2014.

11. Y. Chien, 'Chasing returns has a high cost for investors', Federal Reserve Bank of St Louis study, 14 April 2014.

12. A. Galas, 'Lethargy bordering on sloth: one of Warren Buffett's best investing strategies', *The Motley Fool*, 16 November 2014.

Index